Neurointerventions, Crime, and Punishment

STUDIES IN PENAL THEORY AND PHILOSOPHY

R.A. Duff, Michael Tonry, General Editors

Retributivism Has a Past
Has It a Future?
Edited by Michael Tonry

Punishment, Participatory Democracy, and the Jury
Albert W. Dzur

Just Sentencing
Principles and Procedures for a Workable System
Richard S. Frase

Popular Punishment
On the Normative Significance of Public Opinion
Jesper Ryberg and Julian V. Roberts

Taming the Presumption of Innocence
Richard Lippke

Sentencing Multiple Crimes
Edited by Jesper Ryberg, Julian V. Roberts and Jan W. de Keijser

Playing Fair
Political Obligation and the Problems of Punishment
Richard Dagger

Hate, Politics, Law
Critical Perspectives on Combating Hate
Edited by Thomas Brudholm and Birgitte Schepelern Johansen

Criminal Law in the Age of the Administrative State
Vincent Chiao

Punishment and Citizenship
A Theory of Criminal Disenfranchisement
Milena Tripkovic

Beyond Punishment?
A Normative Account of Collateral Restrictions on Offenders
Zachary Hoskins

Neurointerventions, Crime, and Punishment

Ethical Considerations

JESPER RYBERG

OXFORD
UNIVERSITY PRESS

OXFORD
UNIVERSITY PRESS

Oxford University Press is a department of the University of Oxford. It furthers the University's objective of excellence in research, scholarship, and education by publishing worldwide. Oxford is a registered trade mark of Oxford University Press in the UK and certain other countries.

Published in the United States of America by Oxford University Press
198 Madison Avenue, New York, NY 10016, United States of America.

Library of Congress Cataloging-in-Publication Data
Names: Ryberg, Jesper, author.
Title: Neurointerventions, crime, and punishment : ethical considerations at the intersection of neurolaw and penal theory / Jesper Ryberg.
Description: New York, NY : Oxford University Press, [2019] | Includes index.
Identifiers: LCCN 2019017345 | ISBN 9780190846428 (hardback) |
ISBN 9780190846435 (updf) | ISBN 9780190846442 (epub) | ISBN 9780190846459 (online)
Subjects: LCSH: Criminal behavior. | Criminal psychology. | Crime prevention. | Neurosciences.
Classification: LCC HV6115 .R93 2019 | DDC 365/.6672—dc23
LC record available at https://lccn.loc.gov/2019017345

For
Lisbeth and Flemming

Contents

Acknowledgements

I would like to express my gratitude to all those who have helped and inspired me during my work on this book. My most persistent intellectual debt is to my colleagues in the *Research Group for Criminal Justice Ethics* with whom I have had numerous discussions on the topics of the chapters. Many others have also generously provided me with valuable suggestions and criticisms. Moreover, I am indebted to the Carlsberg Foundation which has provided a perfect setting to carry out the work. Finally, I am very grateful to James Cook of Oxford University Press for his support of the project and to Gillian Rathbone for her indispensable assistance.

Copenhagen, 19 November 2018
Jesper Ryberg

1

Neurointerventions and Crime Prevention

Setting the Stage

Over the last couple of years, an increasing number of philosophers, legal scholars, and criminologists have become preoccupied with the idea of using neurointerventions—that is, interventions which in one way or another operate directly on the brain of a subject—as a possible way of preventing offenders from engaging in future criminal activity. To some, such an idea may sound as a new and shocking example of how scientists can be led astray by following a path that would have been better left untrodden. Whether this is the case, that is, whether the use of neurointerventions on criminals should be regarded as shocking not merely in an immediate emotional way but in a moral sense—as something that is clearly wrong and should be met with vigorous exception—is basically what this book discusses. All the following chapters are devoted to considerations of the ethical aspects surrounding this question. As for the first part of the contention, that the idea is genuinely novel, this is definitely not accurate.

The idea of an association between criminal behaviour and the functioning of the brain is far from new. Even though the term 'criminology' did not appear until the end of the 19th century and the scientific discipline of criminology did not take genuine shape until even later, all the first attempts at understanding the aetiology of crime began with biological theories. As Rafter et al. have outlined in their comprehensive work on *The Criminal Brain*, the earliest origins of criminological thinking attempted to establish how brain impairments might cause criminal behaviour. Nineteenth-century views on crime were heavily influenced by theories of, respectively, the abnormal brain, the atavist brain, and the degenerated brain (Pustilnik 2009; Rafter et al. 2016). Moreover, even though there has been, for both scientific and obvious historical reasons, a significant movement away from the early biological theories in favour of sociological theory, the idea of a biological base for crime has also had a major influence on parts of criminological thinking over the 20th century. In the view of Rafter et al., the 'history of

Neurointerventions, Crime, and Punishment. Jesper Ryberg, Oxford University Press (2020). © Oxford University Press.
DOI: 10.1093/oso/9780190846428.001.0001

biological theories of crime is in many ways the history of criminology itself' (Rafter et al. 2016, p. 10). Even if this may seem an overstatement, it is clear that the idea of a biological base for crime, in which the brain plays the main role, has a long prehistory.

The same is the case with regard to the idea of preventing future crime by influencing the brain. A large part of penological thinking in the last century was heavily influenced by rehabilitationist ideals (e.g. von Hirsch et al. 2009). The rehabilitationist view—that the goal of sentencing is, at least partly, to reform the criminal tendencies of offenders—has been presented in various ways. Some theorists have had their main focus on changing the social causes of crime by emphasizing resocialization and reintegration into the community as the main aims. However, other theorists have defended correctional models of rehabilitationism, according to which the goal is to turn offenders into law-abiding citizens by changing their attitudes and inclinations in more direct ways. The idea of treating or curing criminals has, for a long time, had a significant attraction. Consider the following three questions and statements:

> 'Can new discoveries about the brain reclaim a million criminals? Can psychological research cut America's crime bill in half? Can scientists, using drugs and surgery, eliminate dishonest impulses from the minds of crooks?'

> 'A broad interdisciplinary effort is getting under way to explore the biological nature and origins of violence. Biologists, biochemists, neurophysiologists, geneticists, and other natural scientists are probing with increasingly precise tools and techniques in a field where supposition and speculation have long prevailed. Their work is beginning to provide new clues to the complex ways in which the brain shapes violent behaviour . . . In time, these insights and discoveries could lead to practical action that may inhibit violent acts—perhaps, for example, a change in the way children are brought up, or treatment with 'antiviolence' drugs. Such preventive steps might in the long run be more effective in controlling violent crime than either 'law and order' or social reform.'

> 'Criminal repeat offenders . . . who during medical examination are shown to have neuropathology discernible by the current state of medical knowledge, should, after the appropriate informed consent, be further evaluated

> and treated by qualified medical and nursing personnel . . . Their management should include the usual treatments, for example, removal of brain tumors, control of seizure disorders with anticonvulsants, etc., as well as the administration of psychotropic drugs, ECT, and functional neurosurgery.'[1]

The first passage is taken from a 1939 volume of the magazine *Popular Science* and advances the theory of renowned psychiatrist Carleton Simon—on how a misbalance in the relative dominance of the lobes of the brain may explain the 'bewildering behaviour of many lawbreakers'—and the 'amazing' psychosurgical results of Walter Freeman and James W. Watts (who became the front figures in the lobotomy area in the US). The second passage is from an article published in *Fortune* magazine from 1973 featuring the work of a number of contemporary scientists—including Vernon H. Mark and Frank R. Ervin and their book *Violence and the Brain* (1970)—on how drugs may help in modifying the behaviour of violent prisoners. The final passage is part of a recent overview of the history of psychosurgery, in which professor of neurosurgery Miquel A. Faria suggests that a small group of repeat offenders should undergo medical treatment not by government doctors in public facilities but by physicians in private institutions. Thus, the aims of the three passages differ. However, all clearly demonstrate that the thought of preventing crime through some sort of medical procedure has been advanced in very different epochs and thus is far from new. The idea has been repeatedly defended throughout the 20th century and has—as we will return to in a later chapter—been put into practice in various ways.

But if the idea of using brain-based interventions as an instrument of crime prevention is in itself not a new idea—and, as we return to later, it has often turned out to be associated with naïve science and tragic practice (see Chap. 6)—what then constitutes the background for the recent attention that has been directed to the use of such methods? Why have theorists across different scientific fields, almost as an echo from the past, started considering whether such interventions should be used as an alternative to punishment; whether it is morally acceptable to use penal reductions as incentives to persuade offenders to take part in neurological treatment programmes; or

[1] The first quote is from P. A. Clarkson (1939, p. 45); the second from Bylinsky (1973, p. 134); and the third from Faria (2013, p. 8).

whether neurointerventions can ever be justified as part of mandatory treatment schemes?

Why Consider These Questions?

Though there are always reasons to be somewhat cautious when it comes to the explanation of the causes of trends in the development of ideas in general, and in philosophical thinking in particular—at the individual level there may be various personal motives driving the work of researchers and at a broader level the causes may be multifarious[2]—it is in this case not difficult to identify some obvious causes which at least make it reasonable to engage in considerations of the above questions.

The most important component in the outline of the background for addressing ethical challenges related to the use of neurointerventions as a method of crime prevention, is the fact that an increasing number of scientific studies are being conducted which in one way or another relate to this possibility. It is indisputable that neuroscience has undergone significant developments over the last decades, and that advances in new neuroscientific research tools and technologies have increased the possibility of investigating the human functioning brain. A number of studies have been and are continuously being conducted exploring possible brain dysfunctions in individuals who have demonstrated a propensity for aggression or serious criminal behaviour. These studies have been brought about as a result of the major advances in scanning techniques such as positron emission tomography (PET), single photon emission computed tomography (SPECT) and, not least, functional magnetic resonance imaging (fMRI). In the mid-90s some of the first studies were conducted that examined, for instance, differences in glucose metabolism in the prefrontal cortex across a group of violent offenders and a control group (see Rocque, M. 2013). These early studies have subsequently been followed up by a number of studies using imaging technology to examine brain abnormalities associated with violence. The first meta-analysis was published in 2009, wherein the authors summarize the results of 43 studies and conclude that '[o]verall, findings establish fairly robust and significant prefrontal structural and functional impairments in

[2] For a discussion of some societal trends that have contributed to this discussion, see e.g. Rose (2010).

antisocial population as assessed by brain imaging' (Yang and Raine 2009, p. 86). It is fair to say that these studies are not uncontroversial. But they reflect a movement in the direction of what has become known as *biosocial* criminology which seeks to integrate social and biological causes of crime.[3] What makes this trend relevant in the present context is that the ambition of providing a more coherent picture of the causes of crime by invoking biological considerations is often accompanied by the hope of obtaining new ways of influencing and preventing criminal tendencies in offenders. For instance, a recent study has claimed to demonstrate for the first time in humans that serotonin 1B receptors—called '$5\text{-HT}_{1B}R$'—are related to high levels of anger and aggression in violent offenders but not in control non-offenders. The researchers behind the study conclude on an optimistic note by contending that these receptors may 'represent a molecular target for development of pharmacological antiaggressive treatments' (Cunha-Bang et al. 2016, p. 7). In this way, what seems an obvious step from the study of brain mechanisms predisposed to criminal conduct to suggestions as to the possible remedies, provides the first reason for considering more thoroughly the desirability of using neurointerventions as an instrument of crime prevention. However, other types of research, in a more direct way, provide a ground for engaging in this question.

Many scientific studies, in recent years, have been conducted in order to explore how pharmaceuticals—but also other direct ways of influencing the brain—affect the social behaviour and moral decision-making of humans in general and offenders in particular.[4] To mention a few examples of the more general studies, some selective serotonin reuptake inhibitors (SSRIs), which are usually prescribed for the treatment of depression and anxiety disorders, have in some cases been found to produce changes in the social behaviour of healthy individuals. Participants in a Dictator game who had been administered a particular SSRI (citalopram) were found to divide the sum of money more fairly than members of a control group (Tse and Bond 2002). The same SSRI has also been found to increase aversion to causing harm to others (Crockett et al. 2010). Another example is the neurotransmitter and hormone oxytocin which seems to have an impact on trusting behaviour (Kosfeld et al. 2005; Baumgartner et al. 2008). Experiments in

[3] For considerations on how biosocial criminology has gained acceptance by criminologists, see, for instance, Rafter et al. (2016); Vaske (2017).

[4] For an excellent overview of how a number of pharmaceuticals, which are currently in therapeutic use, may influence moral decision-making and behaviour, see Levy et al. (2014).

which participants play economical games involving trust relations have shown that those participants who were administered oxytocin by nasal spray showed an increased level of trust behaviour. There are also studies indicating that the widely used beta-blocker propranolol may have an effect on moral judgements that are triggered by emotional responses. Generally speaking, there is nothing surprising in the fact that drugs of various sorts may have an impact on the dispositions, motives, decisions, and behaviour of individuals. This insight goes far back in the history of mankind. But it is equally unsurprising that such research has prompted discussions of the possibilities of cognitive, emotional, and even moral enhancements of humans.

Studies have also had a narrower focus on investigating the possibility of influencing the behaviour of offenders. These studies are both correlational and experimental, and they also vary considerably. Some are based on the more general assumption of a negative association between poor nutrition and antisocial and aggressive behaviour. For instance, a study conducted by Schoenthaler et al. examined the effects of vitamin-mineral supplementation on violent and serious antisocial behaviour in a sample of incarcerated juvenile offenders. It was concluded that the 'correction of low blood vitamin concentrations with vitamin-mineral supplements improves brain function and significantly reduces violence among delinquents confined in correctional facilities' (Schoenthaler et al. 1997, p. 343). A study conducted in Aylesbury Prison in the UK found a significant reduction in both violent and non-violent infraction when a group of young adult prisoners were on supplements (Gesch et al. 2002). Along the same lines, a Dutch study has examined the effects of mineral supplements and fatty acids—shown to have an impact on brain structure and function—and reported a significant reduction in incidents observed by the prison staff (Zaalberg et al. 2010). Other studies have reached similar conclusions (see e.g. Choy et al. 2018; Shniderman and Solberg 2015).

Studies have also been carried out on the effects of different types of pharmaceutical drugs. Based on an assumption of an association between serotonergic dysfunction and aggression, an Australian study explored the possibility of using selective SSRIs as a means of controlling impulsive violent behaviour. The researchers observed reductions over a range of behavioural measures (including impulsivity, irritability, anger, and assault) and concluded that the 'findings suggest that treating impulsive violent individuals in the criminal justice system with an SSRI is a potential treatment opportunity for this population' (Butler et al. 2010, p. 1137). A comprehensive Swedish study has examined the use of medication for patients with ADHD (attention deficit-hyperactivity disorder) (Ginsberg and Lindefors 2012). Information was gathered from 25.656 patients with an ADHD diagnosis regarding their pharmacological treatment and criminal convictions over a four-year period. The researchers found that crimes by patients occurred significantly less often during periods in which they were receiving ADHD medication than during periods of non-treatment. Being a correlational study, the inverse association between pharmacological treatment for ADHD and the risk of crime can be given different explanations. But, as the researchers underline, the study is interesting in the light of the fact that the prevalence of ADHD in male inmates is estimated to be between 25 per cent and 40 per cent (see also Lichtenstein et al. 2012).

Almost all studies have explored the use of nutritional supplementation or pharmaceutical drugs. However, a few studies have examined the effects of other techniques—such as tDCS (transcranial direct current stimulation)—on aggression (see Choy et al. 2018). To my knowledge, no studies have been conducted on the possible effects of invasive techniques carried out on offenders. (From a historical perspective—as we shall see later (Chap. 6)—the picture is very different.) However, a few studies have explored the effects of DBS (deep brain stimulation) in patients with Erethism; that is, patients suffering from severe cases of unprovoked aggressive behaviour associated with some degree of mental impairment and gross brain damage. For instance, a Spanish study observed that five out of six patients with pathological aggressiveness 'had a reduction of their outburst of violence of PMH DBS [posteromedial hypothalamic DBS]' (Torres et al. 2013).[5]

Though other studies could be mentioned, the above examples suffice to make it understandable why increasing attention has been directed to ethical discussion surrounding the use of neurointervention as a crime prevention instrument. If researchers are applying new neurotechnological research tools to the study of the biological basis for criminal conduct, and if this is done—at least in some cases—with the explicit intention of paving the way for new methods of preventing crime, then it is reasonable to ask what would

[5] For recent studies involving the use of DBS in patients with severe aggression, see Harat et al. (2015). Here, the researchers conclude that, although the DBS treatment did not bring long-term benefits, it 'freed the patient from aggressive disorders' and 'she regained her drive and joy of life—she participated actively in family and social life' (2014, pp. 310, 315). See also Franzini et al. (2013).

constitute the potential pitfalls and to consider when a move from research in the aetiology of crime to recommended treatment options would be ethically premature. If researchers start investigating the possibility of using drug-based treatment in crime prevention, then it is obviously important to ask whether such treatment is ethically desirable. If improved neuroscientific research methods open the way for studies in the use of invasive techniques as a way of reducing aggression and violence, then it is obviously relevant to ask if it can ever be justified to put such techniques into practice. Thus, in this way, the increased focus on ethical questions pertaining to the use of neurointervention reflects the well-known general picture of how scientific research and the development of new technology often prompt consideration on the desirability of applying the new possibilities as ways of dealing with societal challenges. However, even if the impetus between such ethical considerations at the most basic level is not difficult to understand, the idea of engaging in this ethical discussion may still, as we shall now see, be met with some reluctance.

Does It Make Sense?

Does it make sense to discuss the possibility of using neurointervention as a crime preventive tool, and is such a discussion really desirable? In order to consider these questions and the reservation they express—which is not difficult to grasp in the light of the sordid prehistory of such a treatment paradigm—we will have to clarify the present aim.

As noted above, many theorists have underlined the fact that biological theories of crime have gained a renewed foothold in modern criminology. For instance, as Rafter et al. have put it, the 'history of criminology . . . has been characterized by pendulum swings to, first, biological and then sociological explanations, with the pendulum today hovering indecisively somewhere in between' (Rafter et al. 2016, p. 283). To what extent this picture is accurate and what it more precisely implies, is not something that will be discussed in this book. The complicated discussion of crime causation is one that I will happily leave to criminologists and other theorists who are in that respect much more qualified. Furthermore, with regard to the discussion of crime prevention, there are many methods within a biosocial framework that could in principle be imagined—in particular, in the light of the philosophical prerogative of giving room to hypothetical and futurist scenarios—which will not be subjected to consideration. The discussion in this book will be much more restricted. The purpose here is limited to the ethical discussion of the use of neurointerventions as a method of preventing offenders in engaging in future criminal activity. But this immediately raises the question as to what is meant by neurointervention.

The generally shared view among theorists in this discussion is that neurointerventions cover a range of methods or technologies that affect the conative, affective, or cognitive aspects of the mind of an individual by operating directly on his or her brain.[6] This specification is not precise. For instance, it may not be clear what it means to operate 'directly' on the brain. However, in my view, it is doubtful that a strict definition can be provided and, more importantly, that such a definition is needed. For the present purpose, and in accordance with the way the ethical questions are typically being discussed, it will be sufficient to use 'neurointervention' as a denomination of a range of different techniques including: 1) supplementation and pharmaceutical drugs; 2) non-invasive brain-stimulating techniques, such as tDCS, which works by delivering a low-intensity electrical current to brain areas, or TMS (transcranial magnetic stimulation) in which a magnetic field generator placed outside the head produces small electric currents in brain regions just below the generator; and 3) invasive techniques, such as DBS—which involves the surgical implantation of microelectrodes into a particular region of the brain—or other types of surgery. This specification does not preclude open questions about borderline cases. But none of the arguments in the following hinges on the precise scope of the concept. However, the discussion will be further limited by two presuppositions.

First, it will be assumed that neurointerventions can serve a crime-preventive function by reducing the propensity for crime in at least some individuals who are undergoing treatment. Second, it is assumed that such interventions can be administered without having too severe side-effects on the individual who is being treated. Again, not all the arguments in the following discussion are strictly conditional on these assumptions. But the two assumptions serve a methodological purpose by limiting the scope of the discussion. Without making the first assumption, there will be no discussion. There are, of course, other ways in which neurointerventions could serve functions within the criminal justice system. For instance, there are

[6] From a medical perspective, the term is often defined simply as a treatment approach to conditions that occur within the vessels of the brain or the spinal cavity.

well-known legal discussions of whether it is acceptable to use drugs in order to render a defendant sufficiently competent to stand trial (see e.g. Ryberg 2016a), as well as to restore the competence of incompetent individuals in relation to both pre-trial settings (e.g. competence to make confessions, or participate in line-ups, etc.) and post-trial settings (e.g. competence to motion new trials, parole, and—most bizarrely—to be executed). However, these functions will not be considered here. The focus of the discussion is solely on the use of neurointerventions in relation to crime prevention and this discussion makes little sense if it is not assumed that the interventions could serve such a purpose.

As for the second assumption concerning the side-effects of neurointerventions, this will help in setting the framework of the discussion. If one imagines that a crime's preventive effect could be obtained by administering a neurointervention on an offender without causing any form of adverse effect on this person, then the ethical discussion is almost eliminated. Of course, depending on how the treatment is administered, there could still be questions concerning violations of autonomy, but the ethical challenge is indeed significantly reduced in cases where crimes can be prevented at little or no cost. However, the assumption on an upper level of the side-effects serves the purpose of not turning the discussion into one concerning other traditional and extensively debated issues. It is obvious that neurointerventions can be used to prevent future criminal conduct. For instance, it is possible to administer psychotropic drugs in such a way as to render a person totally incapable of acting, or to perform surgery in ways that will leave an individual in an almost vegetative state. However, if this is what one is considering then the discussion will quickly turn into the traditional dispute in ethical theory as to whether it can ever be justified to sacrifice one person to save others. Moreover, almost to the same effect, the three-drug sequences of the lethal injection cocktail which have become standard procedure in US states practicing the death penalty, can also be regarded as a neurointervention with fatal consequences for the offender to whom it is administered. However, it would side-track what constitutes the more interesting aspects of the discussion of neurointerventions if I had to engage in the already comprehensive discussion on the death penalty and argue why I believe this punishment to be morally wrong. Thus, the assumption regarding side-effects basically serves the purpose of not losing sight of what I believe constitute some of the more important and less debated elements in the current discussion on neurointerventions. However, given this presentation of the overall purpose of the discussion and the delimitations set by the two assumptions, we can now—some might perhaps feel, with even stronger impetus—return to the initial questions regarding the sense and desirability of the whole discussion. Do ethical considerations of the crime preventive use of neurointervention really make sense or is the discussion in itself one that ought to be left untouched and relegated to oblivion? Here are a few reasons for scepticism that may come to mind.

(A) The first reason to question the discussion relates directly to the combination of the two assumptions. By assuming that a neurointervention will in some cases have a crime preventive effect and that this treatment can be administered without too serious side-effects, it may seem as if the discussion has elegantly managed to manoeuvre around what constitutes the main ethical problem, namely, that such treatment will have significant side-effects on persons being treated. In other words, it may seem as if the discussion involves a trivialization by stipulation, in the sense that the significant harm that such interventions may cause, has from the outset been swept under the carpet. However, this objection is not convincing. It may well be the case that the assumption concerning the side-effects will close off some traditional discussions which would otherwise have been in place. In fact, as just noted, this is precisely what motivated the assumption. But this certainly does not make the discussion pointless. I have not excluded the possibility that a treatment may have serious side-effects.[7] And, as we will see in the following, there are many objections that have been raised against the use of neurointerventions that are *not* based on the magnitude of such adverse effects.

(B) Another possible objection, which has some affinity to the previous one, concerns the hypothetical nature of the discussion. It might be felt that the hypothetical character of parts of the ensuing considerations is not an innocent aspect but rather something that brings the whole point of the discussion into question. Consider the following example. An amazing development in the medical treatment of certain neurological diseases is the use of *predictive brain devices*. In some cases, in which a patient suffers from serious incurable epileptic seizures, it has become possible to implant a device which is able to detect oncoming seizures (Gilbert 2015a). The devise will send a warning to the patient when a seizure is about to come on

[7] Given the content of the discussion that follows, there will be no need to engage in more precise considerations of what should be understood by 'too serious' side-effects. As noted, the main purpose has been to rule out extreme cases that would basically amount to a way of sacrificing an individual in the name of crime prevention.

and thereby enable the patient to take the necessary precautions (go inside, lie down, etc.). Moreover, a further possibility is that the detection device sends a signal to an activation device which automatically discharges a drug preventing the seizure. This combined therapeutic system has obvious advantages—for example, that patients are only medicated when it is necessary—and the first detection devices have been successfully implanted. However, the use of such systems has opened up a broader debate as to whether the possibility of detecting and stopping a brain activity, before it manifests itself, could be utilized for dealing with brain disturbances that are not pathologically remarkable but socially undesirable. It has been suggested that this technology could be used as an instrument to monitor and control explosive aggressive outbursts. Suppose, analogously, that a predictive device has been implanted in the brain of an offender suffering from serious problems of impulse control. If the device detected signs of severe aggression—a particular neurological activity above a certain level—it would send a signal to an activation system which would automatically discharge a drug that puts a damper on the activity before it manifests itself in undesirable behaviour. This possibility will no doubt evoke memories in some readers of dystopic literary descriptions of how the minds and behaviour of citizens may be controlled. However, the point here is simply that this discussion is highly hypothetical. A detection and activation system which can react to aggression does not even tentatively constitute a genuine possibility. In fact, it may never become possible. Therefore, could a discussion of this and other like technologies—and perhaps also the assumption concerning the magnitude of the side-effects—rightly be accused of being too unrealistic? Is it reasonable to hold, as Robert Sparrow has suggested in relation to his discussion of so-called moral enhancement (see Chap. 3), that a criteria for the proper use of thought experiments in applied ethics is that such experiments accurately represent and illuminate 'a pressing ethical dilemma' (Sparrow 2014a, p. 113) and that the whole discussion of crime preventive neurointerventions, therefore, should be put to sleep due to its hypothetical character?

There are several reasons as to why an objection along these lines can easily be dismissed. First, the hypothetical nature of the discussion does not necessarily imply that the ethical considerations are at significant risk of being wasted. In applied science, it is sometimes the case that it turns out to be impossible to solve a problem in the way one had imagined and hoped. But this does not necessarily imply that the methods that were developed in the attempt to solve the problem are wasted. Quite often, such methods may find use in relation to other sorts of problem-solving. In my view, the same is often the case in ethics. Thoughts and arguments that are developed in hypothetical contexts may often turn out to applicable and illuminating in relation to other problems, some of which may be genuinely urgent.

Second, and more importantly, the hypothetical nature of part of the discussion that follows in the ensuing chapters is in my view highly desirable. As we have seen, there is ongoing research into the possibility of using neurointerventions in crime prevention. Moreover, as we will return to later, there may exist a political wish or temptation to make use of such treatment options, sometimes even before the effects of the treatment have been sufficiently scientifically validated. Thus, there are strong reasons to engage in considerations of the ethical desirability of such treatment before the technologies are made available and put into practice. In this regard, the present discussion reflects the more general ethical approach to future technology which is often (and reasonably so) based on the assumption that if one does not think of the potential ethical implications of a technology before it is fully developed, one may lose sight of potential desirable applications of the technology or, conversely, be unprepared and overwhelmed by the possibilities of undesirable use. In this regard, the hypothetical nature of the discussion should rather be seen as a virtue.

Third, though the example of the aggression-detecting and brain-modulating device is hypothetical, and even though discussion in the following chapters offers other hypothetical considerations, it is important to underline that a discussion of the use of neurointerventions is not entirely hypothetical. With regard to the assumption concerning side-effects, it is certainly a well-confirmed principle of pharmacology that all pharmaceuticals involve the risk of side-effects. And the effects of invasive techniques are obviously comprehensive. But there are at least a few types of treatment that even today have fewer huge side-effects. For instance, different sorts of nutritional supplementation may have minimal side-effects when administered in proper doses (Choy et al. 2018). More importantly, it is also a fact that some pharmaceutical drugs are already in use in criminal justice practice. As indicated, medication is administered on some prisoners suffering from ADHD. And the most significant example is the use of chemical castration of sex offenders. This type of castration is achieved through the administration of drugs known as MPA (medroxyprogesterone acetate) or CPA (cyproterone acetate) which work by reducing testosterone to a prepubescent level, thereby depriving an individual of most of the capacity to experience sexual desire or engage in sexual activity. Currently, this treatment is authorized in several jurisdictions in the US—California, Florida, Georgia, Louisiana, Montana, Iowa, Wisconsin, and Oregon—and in Europe (Scott and Holmberg 2003; Rice and Harris 2011).[8] A further example that should be mentioned is the use of drugs to combat drug or alcohol addiction in offenders. This group of drugs works in different ways—for instance, by reducing craving in some opioid addicts—and may be administered as part of both mandatory and optional treatment schemes (see e.g. Greely 2008). Therefore, for these reasons, the objection based on the hypothetical nature of the use of neurointerventions in crime prevention does not seem well-founded.

(C) Even though there is no ground for rejecting the significance of an ethical discussion of the use of crime-preventive neurointervention simply because it is hypothetical, there may nevertheless be another reason for scepticism associated with such an in-principle discussion of the *pros* and *cons* of a new technology. It might be objected that these considerations easily end up mis-framing the important discussion of what can and should be done to prevent crime. There is no doubt that there is an immediate attraction in the idea of treating or even curing offenders. As we shall return to later, such an approach to crime and offenders has had a significant influence in penological thinking throughout the 20th century. However, it is equally obvious that there is a major naivety associated with the idea of paralleling criminal misbehaviour and clinical disease. What constitutes a criminal act is a result of political decision-making, and there are some acts which are being criminalized while others are being decriminalized. Furthermore, offenders do not constitute a particular or identifiable subgroup of the population. In reality, the picture is much more blurred. Studies of self-reported offending show that criminal conduct is not an exceptional type of behaviour that can be attributed to a particular group of people. For instance, the *Cambridge Study in Delinquent Developments* found that more than 90% of males in London admitted to having committed at least one crime that could have led to a conviction (Farrington 2002). Though the results of self-reported offending may vary between different studies, such figures certainly lend support to the view that there is some naivety in the idea of preventing crime by curing offenders.

[8] As we will return to at a much later stage, the use of this procedure is highly controversial (Chap. 7)

To the same effect, it might be held that the idea of focusing on neurointervention as a way of preventing crime tends to ignore the fact that crime takes place in a social setting. For instance, this point has been made in relation to the broader ethical discussion that has recently arisen on moral enhancement (see Chap. 3). What critiques have underlined is that, insofar as one promotes the idea of moral enhancement as a solution to urgent societal problems—e.g. war, terrorism, and poverty—one blatantly fails to recognize that wrongful behaviour contributing to the existence of these problems takes place in a societal, political, and economical context that enables and sometimes promotes these behaviours. For such reasons, it has been argued by Inmaculada de Melo-Martin and Arleen Salles that the idea of moral enhancement is 'highly suspect' since it perceives humanity's moral ills as a result of *individual* moral deficits (Melo-Martin and Salles 2014, p. 6). The same criticism could be raised against the idea of preventing crime through brain interventions. The idea of treatment seems to entertain an oversimplified conception of crime as abstracted from the real world. Put simply, it makes little sense to consider treatment of a person who has committed theft or burglary because she lacks access to medicine, is unable to feed her children, or for other reasons directly reflecting social destitution. Focussing on individual neurological or behavioural deficits in such cases would seem to constitute a conspicuous instance of misdiagnosing the problem of crime.

However, even though I am very sympathetic to such considerations, it is in my view a fallacy to present them as an argument against the ethical discussion of the use of neurointerventions. That there are many cases in which such treatment would make little or no sense is hard to deny, even if one holds an extremely optimistic view on what research in this field may accomplish in the future. But this does not show that there could not be *some* cases in which the use of such treatment would be desirable. Be that as it may, the main problem of the above objection is that it is directed against the wrong target. If it is suggested that a neurointervention should be administered on the thief or burglar, then it is certainly reasonable to object and direct attention to the social problems that may have led to the crimes. However, this constitutes an objection against the *method* that is used to handle crime—not against the *discussion* of what constitutes the preferable method. The objection to treatment instead of social intervention or rehabilitation is in itself precisely this: a part of the moral discussion of whether and when the use of such treatment is desirable. Moreover, one might also consider the following question: suppose that research in this field continues to increase and that scientists in the near future develop a range of different drugs and other types of treatment designed to prevent crime and, further, that politicians and other decision-makers are eager to put such treatment into practice. Would it then be preferable not to have had an ethical discussion of what constitute the desirable ways of using these treatment options and what constitute the ethical pitfalls? In my view it is hard to believe the answer to be in the affirmative. Thus, unless one holds otherwise, this indicates that the objection is misdirected.

(D) While the claim that one should not lose sight of—or exclusively focus on—attempts at reducing crime by social means, is in itself a contribution to the discussion as to whether treatment by medical procedures is ethically desirable, there is another reason for reluctance, which directly addresses the desirability of the discussion. It might be held that this discussion in itself may lead politicians and other decision-makes astray. Consider a well-known analogy. A famous argument in the ethical discussion of torture is provided by the ticking-bomb scenario. Suppose that a timed bomb is about to detonate at a school somewhere in the area and that this will cause the deaths of 500 children, but also that the police do not know precisely at which school the bomb is placed. Now, *if* the police knew that a particular terrorist had placed this bomb and that he consequently knew of its location; *if* they knew that the terrorist would provide trustworthy information about the location if he were to be tortured; *if* one knew that this information could not be obtained in other ways and that there would not be time to evacuate all schools in the area; and *if* the use of torture would have no impact on the inclination by the police or other parties to resort to torture in other future cases; then would it be morally acceptable to torture the terrorist in order to gain the requisite information and save the 500 children?

In philosophical discussion, this scenario has been used as a way of examining the plausibility of view that there are absolute deontological constraints against the use of torture. The reason I have italicized the 'ifs' is to underline what philosophers in this discussion fully acknowledge, namely, that this is a highly hypothetical scenario. Without the many stipulations, the scenario would simply not serve the theoretical purpose it is meant to serve. However, some critiques have warned that philosophers should not engage in this sort of discussion at all. In a time of terrorism where it is a sad fact that there is, even among so-called civilized nations, considerable support of various sorts of enhanced interrogation techniques and torture, one should be aware of the risk; namely that politicians might well adopt arguments based on the ticking-bomb scenario and ignore the fact that answers in this hypothetical scenario cannot be transferred to answers as to what would be acceptable in a messy reality. Correspondingly, it might be objected that a hypothetical discussion of whether neurointerventions should ever be used in crime prevention might well be adopted and misused in a political context in which there is little notice taken of the various stipulations underlying the conclusions in the theoretical ethical discussion.

Whether this objection is plausible is ultimately a question of what kind of impact the discussion can be expected to have. Thus, the objection relies heavily on a number of uncertain empirical assumptions, which means that it may be difficult to sustain, but also to definitively rebut. However, here are a few reasons why I am reluctant to accept the objection. First, as we have seen, the ethical discussion arises on the ground of research conducted by physicians and neuroscientists on the possibility of using neurointerventions as a means to influence and counter antisocial behaviour. Whatever one believes about this kind of research, it is a fact that it is being conducted. And, as noted, it is hard to believe that one would be better served in a political context by subtracting the ethical considerations from the discussion. Second, even though part of what will be argued in the following is that some of the in-principle objections to the use of neurointerventions cannot stand closer scrutiny, the arguments to this effect certainly do not amount to anything with the simplicity and immediate appeal as the ticking-bomb scenario. Moreover, in a case where decision-makers need ammunition to rationalize decisions on the use of neurointervention in criminal justice practice, there is no reason to believe that they would not be able to deliver arguments to this effect with much stronger appeal than those that will be presented here. It is a (sad) fact that when it comes to criminal justice issues, politicians already have significant training in powerful ways of selling their message. Third, when it comes to the desirability of implementing the use of neurointervention under circumstances that equal those that currently dominate criminal justice practice, the considerations in the final part of this book end on a rather sceptical note. Surely, this cannot be held to constitute an obvious starting point for political attempts at legitimizing unjustified use of neurointerventions in criminal justice practice? Thus, even though I share the view that the impact of research is in itself susceptible to moral consideration, I do not believe that sufficient has been said to terminate the present discussion before it has begun.

Overview of the Discussion

There are many ways of using neurointerventions on offenders that are clearly wrong and which, precisely for this reason, are ethically uninteresting in the sense that they do not call for any theoretical scrutiny. Analogously, the killing of an innocent person is obviously a tragedy and major problem when it happens but it is not the sort of behaviour which requires thorough ethical consideration. Philosophers and other theorists can of course engage in considerations of *why* precisely such an act is morally wrong but the ethical assessment in itself is not complicated. From any ethical point of view such an act would be wrong. Similarly, it one takes a look at many of the ways in which neurointerventions, throughout the last century, have been administered to offenders in order to prevent recidivism, one can see that there are numerous cases that do not require further ethical consideration; these attempts at treating or curing offenders were wrong *simpliciter*, and do not provoke dispute between people holding different opinions or between ethicists adhering to different theoretical positions. However, it is clear that there are also ways of using neurointerventions that are *prima facie* dilemmatic, which provoke disagreement and dispute, and which therefore call for thorough consideration. The purpose of the first four chapters of this book is to address such issues. What this means is that the chapters are not structured in such a way as to reflect the different types of neurointervention. Even though there are, of course, some neurointerventions which are much more harmful than others to the individual who is being treated, the most controversial ethical challenges do not arise from differences in the types of intervention. For instance, it is not obvious that an invasive neurointervention, to which an offender freely consents, is morally more problematic than a more lenient non-invasive treatment that is forced upon an offender. What this indicates is that the main ethical issues arise from the way in which neurointerventions are being administered. Thus, the first half of the chapters is devoted to discussions of different controversial ways of using this sort of treatment.

Even though there are various examples of the use of neurointervention in crime prevention which are clearly wrong and which, as suggested, are therefore less interesting from a *theoretical* perspective, such cases may nonetheless be interesting for another reason. Precisely because the killing of a person is so clearly wrong, it is certainly an important question whether anything can be done to prevent such wrongful actions taking place. Similarly, even though there are ways of administering neurointerventions that are clearly wrong, it may be interesting to ask why such treatment has nevertheless occurred in various historical periods and, more importantly for the present purpose, what mechanisms might provoke such wrongful use in the future if researchers manage to offer a wider range of treatment options. Though this question opens up a plethora of considerations which cannot all be addressed—and some of which go far beyond standard ethical theorizing—I have nevertheless found it important to devote some discussion to this issue. The final chapters deal with some of the ethical pitfalls related to the real-life use of neurointerventions, and basically reflect the fact that I believe criminal justice practice to be very much at odds with what is ethically desirable. Here comes a brief overview of the more precise questions addressed in each chapter.

Chapter 2 begins with a discussion of the ethical legitimacy of making the sentence of an offender conditional on his or her participation in treatment. This possibility of adjusting a sentence, such as a prison term, to the acceptance of engaging in a treatment programme, can be carried out in various ways. However, the possibility considered in this chapter is that of offering a reduced prison term to offenders who agree to undergo of some sort of neurointervention. Such offers have to some extent already been provided in relation to cases of sex offending, and it is possible to imagine that this procedure could in the future be used as an incentive to accept treatment in relation to other groups of offenders. But is such a procedure ethically acceptable?

The standard objection has been that such treatment is basically coercive. If the alternative to treatment is that an offender has to serve a longer time in prison because he is not given a reduction, then—the argument goes—he is not free to accept the treatment. However, this objection has very often been presented without being followed up by any substantiating considerations of when one can plausibly be held to be free to choose. The main part of the chapter is devoted to this question. Furthermore, it examines other potential objections; to wit, that treatment-conditioned penal reductions constitute either exploitation of the offender or a sort of morally inappropriate offer. The final part of the chapter deals with an aspect of the discussion which has struck me as almost entirely absent in the controversy on 'coercive' treatment, namely the way in which the ethical assessment of such treatment is contingent on penal theoretical considerations.

Chapter 3 concerns the even more controversial question as to whether compulsory use of neurointerventions in the treatment of offenders can ever be morally justified. That there are strong reasons against such use is hard to reject. But does this imply that such treatment is wrong everything considered? And if so, why is this the case? The discussion starts out by examining a number of objections that have been advanced in the broader recent debate on moral enhancement. These objections concern, respectively, the lack of a proper moral yardstick, the freedom of the offender, and the personal identity of the offender. Subsequently, it is considered whether—as several theorists have suggested—it is plausible to hold that the compulsory use of neurointerventions violates an offender's moral right to either physical integrity or mental self-determination.

Chapter 4 addresses a different view on how neurointerventions might be used within the criminal justice system. Retributively minded theorists have constituted some of the main critiques of the use of indeterminate sentencing that dominated in earlier treatment-oriented periods of criminal justice practice. Such theorists might also be sceptical about the use of neurointerventions as a way of dealing with offenders. It might be held that the main purpose of the criminal justice system is to impose a deserved punishment on offenders or at least, if there are other goals, to ensure that offenders are not dealt with in any way that is disproportionate to the gravity of the crime that has been committed. However, this gives rise to a question that, even against the background of the controversial issues dealt with in the second and third chapters, may seem provoking, namely, whether a properly imposed neurointervention can in itself function as a type of punishment. This question opens up a discussion concerning the definition of punishment and, furthermore, a broader discussion as to which requirements should be satisfied for a punishment to be morally justified. Finally, it is considered whether a punishment *qua* neurointervention, as has been suggested by several theorists, would constitute a way of dealing with offenders which is cruel and unusual, or which should, for other reasons, be banned in principle. As such, this chapter is primarily devoted to penal theoretical considerations.

Chapter 5 takes up an issue that more narrowly relates to the question of *who* should administer the use of neurointerventions if such treatment were to be put into practice. That such treatment would usually require at least some degree of involvement by personnel with medical skills is beyond doubt. In fact, in many cases such treatment would require highly specialized medical competences. However, this raises the question as to whether physicians should be involved in the treatment of offenders. More precisely, the purpose of this chapter is to consider whether the combination of the fact that the treatment presupposes medical involvement, and the view that physicians should not be involved in such treatment, is sufficient to block the whole discussion of the morally legitimate use of neurointerventions in crime prevention. It is examined whether any of the arguments, that have been advanced in the discussion of the physician-involvement in the use of lethal injection in relation to capital punishment, manage to establish that physicians have a moral duty to abstain from criminal justice use of treatment by neurointerventions.

Chapter 6 takes its point of departure from the fact that the idea of using neurointervention as an instrument of crime prevention carries a very dark prehistory. If there are readers who, from the outset, feel highly uncomfortable about the overall topic of this book, it is not surprising if the discussion is perceived against a backdrop of historical precedence. However, the history of the use of various types of non-invasive and invasive treatment methods over the last century plays a somewhat unclarified role in the modern neuroethical debate. Some theorists seem almost to believe that the use of neurointerventions constitutes an entire novel ethical issue. Other theorists are much more explicit about the historical precedence but direct attention to how such treatment has previously been administered without devoting much effort to clarifying the present relevance of past experiences. Both approaches are unsatisfactory. Academic amnesia is certainly not a virtue. But it is also clear that previous misconduct and wrongdoing cannot in any simple manner predetermine the answers to present moral problems. Thus, the purpose of this chapter is to consider the moral significance of the sordid prehistory of neurointerventions for the present ethical discussion of the use of such treatment in crime preventive.

Chapter 7 takes the step from traditional abstract ethical considerations to the more practical question of what should be expected if one imagines that treatment in the form of neurointerventions were further developed and were implemented within a framework such as the one we know from current criminal justice practice. The discussion in this chapter is based on the widely accepted view among penal theorists, namely, that current penal practice in the US, and also in many other Western countries, deviates significantly from what is ideally desirable. But if this is the case, what does it mean for the idea of applying neurointerventions within such a practice? This question is examined by reconsidering many of the conclusions that have been drawn in preceding chapters in the light of non-ideal real-life penal practice.

As this brief overview indicates, there are many ethical aspects of the discussion of crime preventive use of neurointerventions that are not being addressed in this book. No single volume can hope to resolve, or even tentatively touch upon, all the complex philosophical problems that emerge at the intersection of neurolaw and penal theory. Though I shall not here try to anticipate the many conclusions that will be drawn in the ensuing chapters, it should be mentioned that, at least, the aspiration has been to show, on the one hand, that many of the in-principle objections that have been raised against the use of neurointerventions are premature; but also, on the other, that given the way criminal justice systems currently operate we may be well-advised not to put such treatment methods into practice.

2
Neurointerventions and Penal Reductions

The development of neurotechnological treatment as a method for preventing recidivism naturally prompts the question how this treatment should be used within the framework of the criminal justice system. The simplest—and obvious—answer is that a treatment scheme should be offered to criminal offenders in cases where it is estimated to have beneficial effects. Treatment schemes of various sorts are already—though with great variation between different jurisdictions—being offered to offenders serving time in prison. Thus, neurointerventions could be offered to offenders in the same way as other treatment programmes (e.g. cognitive behavioural therapy, anger management, etc.), that is, as something the offender can accept or reject but which has no implications with regard to the sentence the offender is serving. Though it is possible to imagine some situations in which such offers would be morally dubious—such as, if offenders are not properly informed about the content of the treatment—it seems reasonable to believe that offers of this type would not usually cause much moral concern. However, the picture seems to change significantly if the treatment is *not* separated from the sentence, that is, if a sentence is in one way or another made conditional on the offender's acceptance of participation in treatment.

There are different ways in which a sentence can be made conditional on treatment participation. For instance, an offender who accepts a treatment may be provided with certain privileges or may be given the opportunity to serve time under less harsh prison conditions. However, in the following, I will focus on the model which has received most attention and which also seems to capture most of the moral concerns that might be raised in relation to the other models, namely, whether it is morally acceptable to offer a more lenient sentence in return for participation in a treatment programme. That is, would it be acceptable to offer a violent offender the

Neurointerventions, Crime, and Punishment. Jesper Ryberg, Oxford University Press (2020). © Oxford University Press.
DOI: 10.1093/oso/9780190846428.001.0001

possibility of serving one year in prison, rather than two years, in return for participation in a pharmacological impulse control treatment programme? Or, as an extreme example, would it be acceptable to significantly reduce the sentence served by a psychopathic offender in return for accepting deep brain stimulation treatment (see Hübner and White 2016; Ryberg 2016b)?

It is not difficult to think of reasons that can motivate this way of connecting sentencing and treatment. There is no doubt that early release from prison would constitute a powerful incentive and that such offers would, compared to models where treatment is only offered as a supplement to a prison term, tempt more offenders to accept the treatment. As Burdon and Gallagher have underlined in their discussion of the treatment of sex offenders: 'one of the most consistently cited findings in studies examining the efficacy of coerced treatment has been its success at bringing clients into and retaining them in treatment' (Burdon and Gallagher 2002, p. 101). Moreover, given the initial assumption that a neurointervention is efficient in terms of crime prevention (without causing too serious side-effects), it may be of benefit to the offender whose risk of falling back into crime and of being re-convicted and re-sentenced is reduced; it may, comparatively, lead to fewer crime victims; and it may save society at large significant costs of keeping offenders behind bars (resources that could be used for various societal purposes).

In this light, it is not surprising that attention has been directed to the possibility of closely connecting sentencing and treatment, and that treatment-conditioned sentencing is to some extent already in use in current criminal justice systems. The two most significant examples are treatment of drug-addicted offenders and sex offenders. With regard to the first group, the idea of coercing offenders into treatment is far from new. In the United States this idea can be traced back to the 'narcotic farms' of the 1920s and 1930s, which treated not only non-criminal volunteers but also offenders who were offered treatment in lieu of punishment. Coercive treatment has later—though to a varying extent and in different forms—been applied in the US criminal justice system. In Australia, coercive treatment of drug offenders has constituted an alternative to incarceration for several decades. And more recently, several other countries have adopted similar initiatives; for instance, Britain, where coercion has become a key element in drug policy (Seddon 2007). With regard to sex offenders, there is also a long tradition of tying sentencing and treatment closely together. While surgical castration has been abandoned as a viable option in most countries, there has been since the late

1990s an increase in the types of method used to pressure sex offenders into other kinds of treatment, including medical castration (see also Chap. 7). In some US states (e.g. Florida) and European countries (e.g. Belgium), chemical castration is offered either as a formal or a *de facto* alternative to further incarceration (Douglas et al. 2013).

The offer of treatment as an alternative to continued imprisonment is therefore, to some extent, already a part of penal practice but it is clear that many regard such procedures as highly problematic. And it is a firm expectation that an expansion of this type of coercive procedure, following the development of new neurotechnological treatment methods, will be regarded by many as no less controversial. Thus, I will not engage in specific discussions of the current use of treatment-conditioned sentencing in this chapter but, more generally, I will consider whether and when this way of using neurointerventions could be morally acceptable.[1] More precisely, we will start by considering whether an offer of penal reductions in return for the acceptance of neurointerventions violates the requirement of informed consent, usually regarded as an imperative in medical treatment. Subsequently, it will be considered whether there is a ground for other types of objections, based on either the exploitative nature or the inappropriateness of such offers. Finally, the issue will be considered from a penal theoretical perspective. It will be argued, on the one hand, that while the view that neurotechnological treatment offers are wrong per se is unpersuasive, on the other hand, there can be penal theoretical reasons against such procedures.

Before turning to the arguments, a single point should be made concerning terminology. When penal reductions are offered in return for participation in treatment, this is usually referred to as 'coercive treatment' (see e.g. Day et al. 2004; Ryberg 2015b). As indicated, part of the reasoning against such offers is based on the view that they are coercive. However, linguistics is not the arbiter of philosophy, and since it seems conceptually messy to consider whether these types of coercive treatment offers are in fact coercive, I will in the following refer to the offer of treatment accompanied by penal reductions as a 'proposed' treatment. Thus, the reader should keep in mind that we are not considering neurointerventions offered simply as a supplement to punishment.

[1] I will not here discuss precisely how a neurointervention should be administered; that is, whether it should be administered during a prison term or after release. Moreover, I will not consider the question whether the need for treatment can be determined at the time of sentencing (for more on this practically urgent issue, see Chap. 7). The question here is whether it is acceptable to offer penal reductions in return for treatment participation.

Do Proposed Neurointerventions Violate the Requirement of Consent?

Consent plays a crucial role in modern medical treatment. As has been underlined, anyone who wishes to argue for forced treatment of competent patients is 'working up a very steep ethical hill' (Caplan 2008, p. 103). In order for a consent to be valid, it has to be informed and freely given. It is the latter condition—the voluntariness clause—that has sometimes been held to be violated in cases where the alternative to treatment is a longer period of incarceration. For instance, in his discussion of castration offered as an alternative to incarceration, Vanderzyl contends that 'the doctrine of informed consent requires a knowledgeable and voluntary decision to undergo treatment, yet offering a convicted offender castration as an alternative to a lengthy prison sentence constitutes an inherently coercive practice rendering truly voluntary consent impossible' (Vanderzyl 1994, p. 140). Other theorists have made the same point.[2] Along the same lines, the more general objection we are considering here is that the use of penal reductions as an alternative to neurointerventions is wrong because this invalidates the consent given by the offender. The argument is based on two premises: that an offender cannot freely consent to a neurointervention when the alternative is continued incarceration; and that valid consent is in fact required in order for a proposed neurointervention to be morally acceptable. Here, we will consider the first premise, while the second will be discussed in the following section.

Is an offender not free to consent to some sort of neurointervention if the alternative is that he will not get a desired penal reduction? Seemingly, a simple question; however, also one that leads into an underlying comprehensive theoretical discussion concerning the relation between offers, threats, coercion, and voluntariness. As we shall see, there are several ways in which a proposal along these lines could be held to be coercive.

A first possibility is to hold that a choice may sometimes place pressure on a person to an extent where the person's control seems to crack under the weight. For instance, Joel Feinberg mentions an example of a person who, as a result of failed brakes, drives his car into a group of children because he is psychologically paralyzed and unable to consider and choose the alternative of swerving

[2] For instance, as part of his discussion of castration, Green contends that: 'Freedom of choice is impossible because the convict's loss of liberty constitutes a deprivation of such a magnitude that he cannot choose freely and voluntarily, but he is forced to give consent to an alternative he would not otherwise have chosen (1986, pp. 16–17).

and going over the edge of a cliff (Feinberg 2003, chap. 23 p. 4). Such cases are most properly characterized as involving psychological compulsion. The person is in no way making a choice. However, in the present context it is clear that this account of involuntariness is not relevant. If an offender is offered either two years in prison or one year with some sort of neurointervention, and if we imagine that the proposal is properly presented and does not require a spur of the moment answer, then it is not plausible to believe that this will usually result in a meltdown of the person's choosing capacities.

However, there is another and more relevant way in which we are sometimes presented with offers that do not seem to leave us with a real choice. If we are placed in circumstances where, given our preferences, one alternative strikes us as uniquely superior to other alternatives, then we might well describe this as a case where we have 'no choice'. Thus, the point is not that our choosing capacity is overwhelmed, but rather that we are in a situation where our sane or rational options are reduced to one. It may be suggested that this way of putting force on the will of a person is precisely what is at stake when treatment is suggested as an alternative to continued imprisonment. Apparently, it is something along these lines which the *Center for Cognitive Liberty and Ethics* (CCLE) has in mind when, in relation to the treatment of drug offenders, it is held that 'being forced to choose between imprisonment or 'medical treatment' with a pharmacotherapy drug is inherently coercive. There are very few things that people will avoid more than going to jail or prison. Informed consent is incompatible with inherently coercive situations that force a person to barter this or her natural neuro- and biochemistry in exchange for freedom' (2004, p. 33). But is it plausible to hold that the prospect of a longer term of incarceration places too much pressure on an offender and, thereby, undermines the possibility of a voluntary consent? I believe that there are two reasons for not accepting this conclusion.

First, it is far from obvious that a choice between alternatives, where one option is uniquely preferable to other options, should be regarded as involuntary. Suppose that a person can choose between her present job or a new job with much better working conditions and a triple salary. Though the person may in ordinary parlance present this as a situation where there is no real choice, it is hardly plausible to maintain that the person who is given this offer cannot really consent to it because the new job is simply too attractive and, consequently, places too much pressure on him or her. To this it might be suggested that coercive pressure is relevant only (or typically) in relation to options we would rather be without, such as the choice between a longer period of imprisonment and a neurointervention. However, if one subscribes to this view, then one will have to explain why the *valence* of the options is relevant for the voluntariness of a choice; that is, why the choice involving a uniquely preferable positive option leaves us free, while the same is not the case if the choice is between options involving a uniquely undesirable negative outcome. Furthermore, even if we are considering a choice between negative outcomes, it is still questionable whether the fact that one option is clearly preferable to another undermines voluntariness. Suppose a medical doctor says to a patient: 'You can choose either between our standard operation procedure which is quite painful and has several other serious side-effects, or our brand-new procedure which has almost no side-effects. Which do you choose?'. The question might of course seem odd simply because the choice is so obvious; however, it is hardly plausible to hold that the patient's choice is *involuntary* and that she cannot validly consent to the new procedure. In his analysis of 'no choice' situations, Wertheimer makes a similar point when he contends that if 'B agrees to an amputation of his leg in order to avoid certain death, he cannot later sue for battery, on the grounds that his consent was not valid because he had no choice' (Wertheimer 1987, p. 197). Even if one can make *some* sense of the idea of having 'no choice' by reference to alternatives and a person's preferences, according to Wertheimer, such situations carry no moral significance.

Second, even if one believes that the above considerations are somewhat premature, and that there is more to the idea that the nature of the alternative in itself implies that there is a choice but no real choosing, one will still have to admit that the question of the voluntariness of the choice must be a matter of degree.[3] But this seems to imply that there are at least some degrees of involuntariness which are not sufficient to render a consent invalid. Suppose again, for instance, that one believes that a person, as a result of the distance in the preferences between the options, has no real choice. In this case, one can try to modify the choice to options that are not too distant.[4] If an offender is offered the choice between the full prison term or a reduced term

[3] For a comprehensive discussion of what determines the degree of the force on a person's will, see Feinberg (2003, chap. 23).

[4] That the degree of force on a person's will depends upon the distance between the options, is only one possible approach. Feinberg considers several other approaches to determining the degree of this force, though ultimately without subscribing to a particular model (see Feinberg 2003, chap. 23). However, even if one believes that the degree should be determined in other ways, the challenge basically remains the same: first, one will have to establish how powerful the force on the will should be in order for it to make a consent invalid; second, it will have to be shown that cases where penal reductions are offered as an incentive for accepting neurointerventions cannot lie below this limit.

supplied by a neurointervention, then one will just have to make sure that the latter option is not manifoldly preferable to the former. In other words, it seems that one can organize the offer of neurointerventions in return for penal reductions in such a manner that the force on the will of the offender is insufficient to invalidate the offender's consent. Thus, in sum, it is questionable whether the use of penal reductions as incentives for offenders to accept neurointerventions can plausibly be held to undermine voluntariness and, even if this is sometimes the case, it still seems that some offers of penal reductions and neurointerventions can be designed in a way that leaves the offender sufficiently free to give a valid consent.

But what follows if these considerations are correct? The arguments may question the plausibility of the view that consent is made impossible because the prospect of penal reductions places too much force on the will of an offender to satisfy the voluntariness clause for valid consent. However, this does not settle the question as to whether proposed neurointerventions rule out the possibility of giving consent. It may be held that the coerciveness does not only concern the comparison—or 'subjective price tag' (Feinberg 2003, chap. 23)—of the proposed alternatives but, more broadly, the circumstances under which a proposal is made. Feinberg has an example illustrating this point. Suppose that Mother Theresa is told 'Give me your money or I will blow your brains out' and that Mother Theresa was about to offer money to the threatening person anyway, out of charity. In that case, it does not seem that Mother Theresa acts *from* coercion when she gives her money—she is acting willingly—but it could still be held that her act is involuntary in the sense that it is done *under* coercion (Feinberg 2003, chap. 23 p. 25). Thus, when we consider how free a person is to act otherwise than the way in which he does, we will have to consider 'the whole coercive burden' under which the person is acting (p. 25). This opens up a set of considerations that has received considerable attention in the modern discussion, namely, what the relation is between coercion and respectively threats and offers. Let us see how some of the views in this classical controversy may affect our discussion of proposed neurointerventions.

To contend that all or most offers involve an element of coercion is clearly implausible. A person is not coerced and left without the possibility of genuinely consenting if she is offered either tea or coffee, or if she is given the choice between her present job or a new one. On the other hand, the classical gunman proposal 'your money or your life' seems equally clearly to involve coercion. Thus, when should proposals be regarded as coercive and when is this not the case? Though, as we shall see, views diverge somewhat on this issue, the standard position has been that (some) threats involve coercion while offers do not. It might perhaps be asked what the precise relation is between coercion, voluntariness, and consent. However, let us for the sake of the argument—and in accordance with the standard view—assume that coercion involved in threats does undermine valid consent (see Wertheimer 1987, chap. 16; Anderson 2011). The question then arises: Do proposed neurointerventions constitute an offer or a threat? Is the state merely offering an offender a neurointervention and an accompanying penal reduction, or should it rather be regarded as a threat when the alternative is no penal reduction, that is, longer time served behind bars?

The answer as to how threats and offers should be distinguished is far from simple. Roughly put, it seems that a threat is unwelcome in a way that an offer is not. More precisely, following the thoughts presented in Nozick's seminal work on the issue, we can say that someone threatens a person by proposing to make the person worse-off relative to some baseline; whereas someone makes an offer to a person by proposing to make the person better-off relative to some baseline (Wertheimer 1987 p. 204; Feinberg 2003, chap 23 p. 40; Anderson 2011, p. 10). If I am given the opportunity of a new job, then this will usually constitute an offer. (If I do not accept it, I am in no worse a situation than I would have been in had I not been given this possibility). However, the choice set up by the gunman, that of losing either my life or my money, surely presents me with alternatives that are worse than the situation I would have been in had he not interfered in my life. Thus, on the suggested definition, the former is an offer, while the latter is a threat. This seems highly plausible. However, the problem is that on closer scrutiny it is not clear how the relevant baseline should be determined. What Nozick suggested is that the baseline for the comparison is determined by what would have been the case in 'the normal or natural or expected course of events' (1969, p. 447). However, as he also pointed out, this specification of a baseline in itself opens up several interpretations. It may be interpreted descriptively (predictively or non-normatively) or it may be interpreted in a normative manner. The former concerns what would have been the case had the communication not taken place, while the latter concerns what *should* have been the alternative. In fact, the discussion has shown that the descriptive approach in itself is open to several interpretations (see e.g. Feinberg

2003, chap. 23; Wertheimer 1987, chap. 12). The point is of course that the different ways of determining the baseline may leave us with varying answers as to whether a bi-conditional proposal is an offer or a threat. Then, which interpretation is the right one?

At this point, there has been a major disagreement between theorists and I shall not try to settle the question here. In fact, as Wertheimer has underlined, given the contextual nature of coercive claims, it need not be the case that there is always only one interpretation of the baseline that takes precedence over other interpretations (Wertheimer 1987, p. 212f). Thus, rather than considering the distinction between coercive and non-coercive proposals in general terms, it will be more helpful to narrowly consider what different interpretations imply specifically in relation to the proposal of neurointervention in relation to offenders.

Suppose first that we follow a non-normative interpretation of a baseline by asking what would be the case if the proposal of accepting some sort of neurointervention had not been made to the offender. It seems that a plausible answer is that the offender would receive the prison term that was warranted on the ground of the crime he had committed. This would be the *normal or expected* course of events, given the fact that the person had been convicted for a crime and is being dealt with within the framework of the criminal justice system. Thus, on this interpretation the proposal constitutes an offer and, hence, is not a coercive interference (see e.g. Bomann-Larsen 2013). Suppose, alternatively, that we adopt a normative interpretation of the baseline. In that case, the question is what *should* be the course of events that would follow if the proposal is not made to the offender? Again, the answer seems straightforward: The offender should be punished for the crime he has committed. Thus, also on this interpretation, we reach the answer that the proposed neurointervention is *offered* and does not involve coercion.

There is, however, one important condition that must be satisfied, namely, that the punishment which is imposed on the offender is morally justified. If the punishment the offender will have to go through if the proposal is not accepted, is too severe—for instance, if the offender is simply kept in prison indeterminately until he at some point accepts the treatment—then the proposal will no longer be an offer but a threat. Thus, whether the proposed neurointervention constitutes an offer or a threat ultimately depends upon how severely different crimes should be punished and, therefore, on what constitutes the most plausible theory of punishment (see Chap. 7). However, there is no need to consider this question any further. What matters is that, on this account, neurointerventions can be proposed as an offer.[5] Therefore, on both a non-normative and a normative interpretation of the baseline, it seems that neurointerventions can be genuinely offered to offenders as alternatives to continued incarceration, and that this will not undermine the possibility of obtaining valid consent. As indicated, though, the two suggested interpretations do not necessarily cover all possible interpretations of baselines. It should be noted however that, if one wishes to suggest that a proposed neurointervention constitutes a threat, then it seems that one will have to rely on an interpretation according to which, the fact that the criminal justice system could decide to release the offender should be included in the baseline. However, since this is neither what would usually be the case nor what is morally desirable (if the punishment is justified), it is hard to believe that such an interpretation could be plausibly sustained.

The previous considerations draw on the standard view, that the distinction between coercive and non-coercive proposals follows the line between offers and threats. But is this view too restrictive? Could it be the case that a proposed neurointervention accompanied by a penal reduction is coercive simply because, even though it is most properly characterized as an offer, there are some offers which in themselves are coercive? Some theorists believe that coercive offers do exist. At first sight, this may sound a dubious position. How can an offer that adds a new option to a set of options be coercive? However, Feinberg believes that there exist what he refers to as 'freedom-enhancing coercive offers' (2003, chap. 24); namely, situations where a proposer forces another person to do what the proposer wants by manipulating the options in such a way as to render 'alternative choices ineligible' (p. 6). Consider the following example: a kidnapper brings a person to an island where the kidnapper's factory is located, and abandons the person on the beach. The next day the kidnapper approaches the person with the proposal 'Take a job in my factory and I won't let you starve' (Feinberg 2003, chap. 24 p. 7; Zimmerman 1981). According to Feinberg this proposal does not constitute a threat, but a coercive offer. More precisely, he suggests that such offers are characterized by, on the one hand, the fact that there is a

[5] Wertheimer defends this approach. He considers a real case in which a judge made the following proposal to a couple convicted of the death of their daughter by criminal neglect: Two and a half years in prison or one and a half years in prison if you agree to sterilization. As Wertheimer rightly notes, this proposal may be objectionable on a variety of grounds. However, his point is that if the punishment (the two and a half years) is just, then the judge's proposal constitutes a 'noncoercive offer' (1987, p. 218).

significant value gap between the offered alternatives, and that one of the alternatives constitutes an 'intolerable evil' (Feinberg 2003, chap. 24 p. 8). On the other hand, the proposer has arranged the circumstances under which the offer is presented. In the case of the kidnapper, it is clear that the prospect of starvation constitutes an intolerable evil compared to the possibility of working at the factory. Moreover, it is clear that the choice is made on grounds of the fact that the kidnapper brought the person to the island in the first place.

Feinberg's account of coercive offers is far from uncontroversial (see e.g. Wertheimer 1987, chap. 13). However, let us *arguendo* assume that the account is plausible. What, then, does this imply with regard to the proposal of neurointerventions for offenders? Can this proposal be regarded as a coercive offer in such a way as to vitiate the offender's consent? The answer must be in the negative. First, if one, as does Feinberg, holds that it is a precondition that in order for an offer to be coercive there exists—from the perspective of the recipient's scale of preferences—a large distance or gap between the offered alternatives, then this can easily be avoided. As we have seen above, the size of the penal reduction can be adjusted in such a way as to leave only a small distance between the alternatives of continued imprisonment and neurointervention/penal reduction. Moreover, it can perhaps be discussed to what extent a longer rather than a shorter time served in prison constitutes an 'intolerable evil'. However, even if we leave these points out of consideration, the second part of Feinberg's proposal is sufficient to establish that a proposed neurointervention does not constitute a coercive offer. Feinberg's idea seems to be that offers where the circumstances are 'deliberately created by the offerer for the purposes of exploitation do normally invalidate consent' (Feinberg 2003, chap. 24 p. 24). However, this cannot plausibly be held to be the case when neurointerventions are offered within the criminal justice system.[6] It is true, of course, that the criminal justice system has placed the offender behind bars. However, unlike the kidnapper who forces a person onto an island in order to get a new employee working in his factory, the state has not imprisoned the offender with the purpose of urging him to accept a

[6] To my knowledge, Feinberg does not explicitly consider how consent is affected in cases in which an offerer has arranged the circumstances, but has done so for other (good) reasons than to manipulate the ground on which the recipient is given an offer (such as is the case in a criminal justice system that imposes morally justified punishments on offenders). However, it seems reasonable to believe that he would interpret these cases on a par with cases where a party opportunistically uses the circumstances of a vulnerable person as a background for an offer. In such cases, he claims that voluntariness is not reduced to a level that renders consent invalid (see e.g. Feinberg 2003, chap. 24 p. 26).

neurointervention. The state has sent the offender to prison because this is the ethically required response to the offender's misdeed.[7] Thus, even if one accepts Feinberg's controversial idea that some offers can be coercive, it still seems that a scheme involving proposed neurointerventions to an offender can be put into practice in a way that does not invalidate the offender's consent.

In summary, what we have seen in this section is that, despite some critiques, it is far from clear that the use of penal reductions as an incentive to get offenders to accept some sort of neurointervention can plausibly be blocked on the ground that the offender's consent to the treatment would be invalid. It is not plausible to expect that proposed neurointerventions will cause a genuine breakdown of the offender's choosing capacities. And it seems that such proposals can be presented without placing too much coercive force on the will of the offender. Moreover, it has been argued that, on several accounts of the distinction, it seems to follow that proposed neurointerventions do not constitute threats but rather genuine offers. Finally, we have seen that, even if one accepts the controversial idea that there exist coercive offers, proposed neurointerventions do not seem to fall into this category. However, we have also seen that, at least on some interpretations of the contents of threats and offers, it is a necessary condition that the punishment inflicted on an offender (if the proposed neurointervention is not accepted) is morally justified. We will return to this precondition shortly (and in Chap. 7). Finally, it should be mentioned that the previous discussion does not provide a complete overview of all possible theories that have been presented of what distinguishes threats and offers and, more generally, of what characterizes coercion. However, as we shall now see, there is another and more general way in which the consent-based objection to proposed neurointerventions can be challenged.

Is Valid Consent a Necessary Requirement?

The whole point of considering whether proposed neurointerventions are coercive is, as we have seen, the belief that coercion undermines voluntariness and, thereby, excludes the possibility of giving a valid consent. The

[7] The point, of course, is not to suggest that state punishments are always justified. In fact (we shall return to this later), there are strong reasons to believe that this is often not the case (Chap. 7). The point here is simply that, if the offender is punished because this is the morally right reaction to the crime, then Feinberg's condition for an offer to be coercive is not satisfied.

objection is premised on the assumption that valid consent is a necessary condition for a proposed neurointervention to be morally acceptable (or, at least, that there is something wrong in administering neurointerventions on offenders without valid consent). But should we accept this basic premise?

At first glance, the reason for claiming that neurointerventions require informed consent is pretty obvious. No matter whether we are considering the use of pharmacology, non-invasive techniques, or brain surgery, it seems that neurointerventions constitute a type of medical intervention and, as is the case for other types of treatment of competent persons, therefore, presuppose that the patient accepts the intervention. In this respect, the use of neurointerventions we are here considering does not differ from other types of treatment that could be offered to offenders. If an offender suffers from a disease, he will (or at least should) be free to refuse medical treatment. The same is the case with regard to neurointerventions. However, on closer scrutiny, this reasoning may not be as obvious as it appears. Before we see why such is the case, it is important to underline what it implies to challenge the consent requirement. To suggest that the use of neurointerventions for crime preventive reasons does not presuppose a valid consent is not tantamount to holding that it is justified to use such interventions as part of a compulsory treatment scheme. Many theorists regard compulsory neurointervention as wrong for a variety of different reasons—we will return to this in the following chapter. Furthermore, the rejection of the consent requirement does not imply that there is no moral difference between proposing a neurointervention to an offender under coercive circumstances and forcing an intervention onto the offender. There could be several reasons for preferring the former approach, even if the person is not free to decide in either case. All one is claiming, if one rejects the consent requirement, is that the fact that the conditions for consent are not satisfied does not contribute to the wrongness of proposing or using such interventions. Let us see what could possibly sustain this view.

Even though a requirement of consent can, and has been, defended in various ways, a typical claim is that the importance of consent reflects the moral significance of autonomy. However, even if one accepts that one should usually respect a patient's autonomous decision when it comes to medical treatment, it does not necessarily follow that the same is the case with regard to the use of neurointerventions on offenders. It could be held that, even though standard medical treatment can have consequences for parties other than the patient himself (e.g. the patient's relatives), it is usually the case that the acceptance or refusal of a particular treatment primarily affects the life of the patient. But the same is usually not the case if one is considering crime preventive neurointerventions. If a neurointervention implies that other people will not become crime victims, then acceptance or refusal of the intervention is something that has serious consequences for other parties. And this could be held to weaken the weight of autonomy. In fact, the same is sometimes seen even in traditional medical treatment.[8] In some countries, coercive measures are used to force people into accepting vaccinations against serious diseases.[9] And the reason is, of course, that the refusal of vaccination may have serious consequences for other people. As an extreme case, the health-care system sometimes uses forcible quarantine as a way of preventing a patient with an infectious disease from spreading this disease to others. In short, it could be held that it is important that people can make decisions regarding their own lives but, at the same time, that the significance of autonomy decreases when the lives of other parties could be seriously affected.[10]

Another reason for holding that respect for autonomy, as manifested in the consent requirement, cannot necessarily be extrapolated from standard medical treatment to the use of neurointerventions on offenders is that the latter type of intervention may in itself have consequences with regard to autonomy. More precisely, it has been suggested that neurointerventions may sometimes enable or enhance autonomy and that, when this is the case, respect for autonomy actually speaks *in favour* of neurointerventions even if such interventions are proposed under circumstances that are coercive and therefore do not allow for the possibility of valid consent being obtained. For instance, Douglas et al. have argued that medical castration offered to a sex offender may, if accepted, increase the offender's future autonomy by 'removing internal, psychological barriers to autonomy such as irresistible sexual urges' (Douglas et al. 2013, p. 403).[11] A similar view has been defended

[8] Even though the right to reject treatment is usually regard as highly important, it is also a fact that this right is often not regarded as absolute. For examples of circumstances in which forced treatment is considered ethical, see e.g. Focquaert (2014).

[9] In some countries (e.g. Poland and Belgium) parents are fined if they do not comply with vaccination programmes. In other countries (e.g. the US and Spain) certain vaccinations are a precondition for enrolment for school.

[10] For an in-depth discussion of the moral comparison of medical quarantine and non-consensual criminal rehabilitation, see Pugh and Douglas (2016a).

[11] In fact, Douglas et al. argue that offering castration to sex offenders may not only increase the future autonomy of these offenders, but also the offenders' *present* autonomy by increasing the number of alternatives open to these persons.

in relation to the treatment of drug offenders (see Caplan 2006) and could perhaps also be made with regard to interventions assisting offenders suffering from a lack of impulse control. Thus, in this view, respect for autonomy has been presented as an *objection* to the consent requirement (see Pugh and Douglas 2016b).

Even though I believe there is something in both the above arguments, they probably do not succeed in convincing all those theorists who believe that consent should be regarded as morally crucial. No doubt, some will maintain that neurointerventions do constitute a type of treatment and, as such, should not be exempted from the consent requirement. And that, even if a person's future autonomy could be enhanced, there will still exist a constraint against violating the present autonomy—as is the case if valid consent is not obtained. However, there is yet another reason for doubting that the consent requirement must be satisfied in the circumstances we are here considering. It should be kept in mind that a punitive system constitutes the quintessential example of a compulsory context. No one appears to hold that a proper imposition of a punishment presupposes consent by the offender. Thus why, then, should the imposition of a neurointervention presuppose consent? There seems to be a tension between holding, on the one hand, that it is acceptable to use a severe compulsory measure such as putting an offender behind bars while, on the other hand, insisting that the imposition of a neurointervention presupposes informed consent. As pointed out by Pugh and Douglas, this tension confronts the adherent to the consent requirement with the challenge of giving an account of the moral difference between incarcerating an offender without consent and imposing a neurointervention (Pugh and Douglas 2016b, p. 105). On closer scrutiny, it turns out that this task may not be as simple as one might perhaps *prima facie* expect.

The standard view, that consent is a way of respecting autonomy, is of little help. Autonomous decision-making is certainly not what is at stake when an offender is imprisoned. Perhaps it could be held that the difference is that, while incarceration of an offender is morally justified, this is not the case with regard to neurointerventions. However, this view—to which we will turn in the next section—cannot be taken for granted. Presumably, as already indicated, the most obvious answer is that, while incarceration constitutes a punishment, this is not the case with regard to a neurointervention. Neurointervention is a medical treatment and, as such, requires consent. However, this answer is not satisfactory.

First, it should be noted that this answer is not open to all theorists. For instance, some theorists hold that a procedure that increases a person's impulse control should be regarded rather as an enhancement than a treatment.[12] Second, and in my view more importantly, it is far from obvious that the mere classification carries any moral significance. It might well be held that, even though it is usually important to respect autonomous decision-making in relation to medical treatment, this does not in itself establish that there do not exist some subcategories of treatment where it is less clear that autonomy should be respected. For instance, as we have seen, this could be the case where the treatment does not exclusively or primarily affect an individual himself but may have serious consequences for others. Third, even if one is not persuaded by the former arguments, there is a final possibility that would definitely eliminate the requirement of consent in relation to neurointerventions. If a neurointervention could be imposed on an offender, not as a treatment presented as an alternative to punishment but rather as a punishment in itself, then the consent requirement cannot plausibly be maintained. Needless to say, the fact that neurointerventions involve techniques that are traditionally used for medical purposes does not exclude the possibility that the same techniques can be used in ways that do not constitute medical treatment (see Chap. 6). Thus, if a neurointervention can function as, and is imposed as, a punishment, then it seems to follow that the consent requirement can no longer be upheld; that is, at least, if one subscribes to the standard view that punishment of offenders does not presuppose consent. No doubt, the view that a neurointervention could work as a punishment is highly controversial and it may strike some as almost absurd. However, as we will see in Chapter 4, on closer scrutiny there is much to say in favour of this outlook. For the present, it is sufficient to conclude that there are several reasons to question the requirement of consent as a necessary condition for the morally legitimate use of neurointerventions, and that there is at least one reason which—if it can be sustained—is sufficient to definitively reject this requirement.[13]

[12] I will not here enter the discussion about the treatment/enhancement distinction, except by the remarking that I am somewhat sceptical about it. Thus, as indicated, I regard the second reply as the one that is important. For a discussion of the treatment/enhancement distinction, see, for instance, Holtug (2009).

[13] It might perhaps be held that there is another reason for maintaining that neurointerventions require consent, namely, that if one does not obtain consent from offenders, this will ultimately undermine the more general respect for autonomy and consent in other contexts; such as, in traditional medical healthcare. However, the problem is that this claim is hard to sustain. For a discussion of this slippery slope argument, see Pugh and Douglas (2016b).

Exploitative and Inappropriate Offers

Suppose that the previous considerations are sufficient to reject the consent-based objection to proposed neurointerventions. That is, suppose that consent is not a necessary condition for legitimate interventions in a criminal-justice context or, even if it is a necessary condition, that offenders are free to give their valid consent. Does this imply that there is no reason for concern when it comes to offering neurointerventions as an alternative to serving full time in prison? The answer is in the negative. There could be other reasons that do not hinge on the question as to whether an offender's choice is coercive. In this section, we will consider two such reasons. The first concerns whether proposed neurointerventions involve an element of wrongful exploitation. The second holds that at least some proposed neurointerventions belong to a category of offers which, even though they can be validly consented to, are nonetheless inappropriate for the state to make. Let us consider the two objections in turn.

Can the offer of penal reductions in return for accepting participation in treatment using neurointerventions be regarded as a way of exploiting offenders and the circumstances they are living under when placed behind bars? This question has not received much attention. However, there are various other contexts in which the concern for exploitation has been addressed (e.g. economic transactions, use of labour force, trafficking, etc.). Consider the following example presented by Mikhail Valdman: B is hiking in a remote forest and is bitten by a rare poisonous snake. His death is imminent. However, fortunately, another hiker A comes by with an antidote and offers to sell it to B. Though the retail price is $10, A offers to sell B the antidote for no less than $20,000. B would rather lose his money than his life, and accepts the offer (Valdman 2009, p. 3). Even though this example may perhaps strike some as over the top it nicely illustrates what is usually regarded as the more general feature of exploitation, namely, that someone takes advantage of the fact that another party is placed in dire circumstances.

There are two things worth noticing about the idea of exploitation that make the concept potentially interesting in the present context. First, exploitation need not involve any harm to the individual who is being exploited. Obviously, there can be cases in which exploitation advances the interests of one party while harming those of another (e.g. slavery) but, as the example illustrates, this need not be the case. Relative to a baseline where the hiker who was bitten by the snake is not offered the antidote, it is clear that he has benefited (he did not lose his life). Thus, a transaction can be mutually beneficial but still exploitive. The fact that offenders, as we have seen, may benefit in various ways from accepting participation in a treatment scheme does not, therefore, a priori exclude the possibility that the proposal involves exploitation. Second, exponents of theories of exploitation typically hold that exploitation need not involve any sort of coercion. Thus, even though it is reasonable to believe that the hiker who has been bitten by the snake is under intense pressure, this is not necessarily sufficient to undermine consent. The transaction (at least, according to some interpretations) is not a threat and not a coercive offer of the type where a person's conditions of choice have been manipulated by the offerer (see e.g. Valdman 2009).[14] Therefore, if it is true that proposed neurointerventions to offenders are not coercive, there is still the possibility that such proposals are vulnerable to objections concerning exploitation.

However, it is also clear that there are many kinds of advantage-taking that cannot possibly be described as exploitative. For instance, an employer may well benefit from offering a poor person a job, but this does not turn the offer into an exploitative transaction. Moreover, there are several situations where it makes sense to talk of exploitation, but which do not carry any moral significance. For instance, a chess player may be held to exploit a weakness in the opponent's defence (Zwolinski and Wertheimer 2016). Thus, it is clear that further explication of the idea of wrongful exploitation is required in order to examine whether proposed neurointerventions constitute a problem in this respect. However, at this point there is little consensus.

Consider first the so-called fairness account. According to this theory, exploitation is basically a question of taking unfair advantage; that is, A can be said to exploit B when A takes unfair advantage of B (see e.g. Wertheimer 1996, p. 10). The unfairness may come about in one of two ways. It may refer to a defect in the process in the transaction between A and B, such as, for instance, if B's free choice is compromised by misinformation, coercion or

[14] If one subscribes to the view that threats are defined in terms of a normative baseline, it might perhaps be held that the proposal of the significantly over-priced antidote does constitute a threat: the proper baseline is that the hiker who possesses the antidote should have offered this to the bitten hiker for free (or at least for the retail price). However, as argued by Valdman, the wrongness in the antidote case is not well explained merely by referring to the coercive nature of the choice. Consider, as a parallel example, that B is about to drown in rough and dangerous waters, and that A is only a mediocre swimmer. Suppose further, that A nevertheless offers to rescue B if B pays him an excessive amount of money. This case equals the antidote case in important respects but here it cannot be held that there is coercion involved. Given the fact that A, under the circumstances, is not morally obliged to jump into the water, there is no moral baseline implying that the proposal is coercive. But, as Valdman suggests, the proposal still seems exploitive (2009, p. 4).

fraud. In relation to proposed neurointerventions to the offender, these procedural defects need not constitute a problem. The other possibility is that unfairness relates to the *outcome* of the transaction. Even if B benefits from the transaction, relative to a non-transaction baseline, it could still be said that a transaction is exploitative if B does not benefit as much as he ought to relative to some normative baseline or, more generally, if the outcome results in an unfair benefit profile. Neither does this account, however, seem to reflect what is involved in a proposal of neurointerventions to offenders. While it seems plausible to contend that there is something unfair in the benefit profile, when the bitten hiker is paying $20.000 for an antidote at a retail price of $10, it is much less obvious that this is the case when an offender is offered two years in prison rather than four years if he agrees to attend a treatment scheme. Moreover, if we imagine that the theory has been developed so that it is made clear precisely how an unfair benefit should be determined, there is no reason to believe that the size of the benefit cannot be accounted for in the proposal made to an offender (e.g. by appropriately adjusting the size of the penal reduction to the size of whatever is regarded as the benefit of the state or the criminal justice system by offering the treatment).

A different approach to exploitation has been to pay direct attention to the circumstances under which someone is placed when he agrees to engage in a transaction. It might be held that what is essential about exploitation is not merely that someone profits unfairly from a certain transaction, but that one takes advantage of the unfortunate circumstances in which the other party to the transaction happens to be. This is clearly illustrated in the antidote case, where the bitten hiker obviously would never have dreamt of paying such an exorbitant price for an antidote had it not been that without it he would lose his life. Following this line of thought, Valdman has argued that the wrongness of wrongful exploitation is best explained by our obligation not to extract excessive benefits from those who cannot reasonably refuse an offer (Valdman 2009). A somewhat different account, also emphasizing the significance of the circumstance of the exploitative offer, has been presented by Goodin. In his view, exploitation does not concern a specific end-state but consists in the process of 'playing for advantage in situations where it is inappropriate to do so' (Goodin 1987, p. 184). An occasion of exploitation arises when a party is in a particularly strong position vis-à-vis another, and the stronger part positively plays for his own advantage against the other party whom he is 'duty-bound not only not to press advantage against but actually to protect' (Goodin 1987, p. 188).

Applying such circumstance-based accounts to the question of proposed neurointerventions to offenders may seem to have the advantage of reminding us that the treatment here is being offered to a particular group of people who are in fact placed in highly uncomfortable circumstances (namely, in prison). However, apart from this, I believe it is clear that these accounts of exploitation have little to offer in the present context. First, it is not at all apparent that what the criminal justice system or the state gain from offering neurointerventions to offenders constitutes (if accepted) an 'excessive' gain or that it is not possible to adjust to this requirement once a more precise guideline for excessiveness has been specified. Second, with regard to Goodin's account, he specifically underlines that what it means to play for an unfair advantage is contextually determined. And it is far from clear that what is taking place when an offender is offered a treatment can properly be characterized as 'playing for an unfair advantage'.[15] It certainly makes perfect sense to hold that there is a duty to assist the hiker in the antidote case rather than take advantage of his terrible situation. However, when it comes to proposed neurointerventions to offenders, there is no corresponding call for assistance. In so far as the sentencing of an offender is morally required, the alternative to a proposed treatment is not assistance, that is, an attempt to help the offender out of some sort of fatal situation. Rather is it to ensure that the appropriate sentence is imposed upon him, that is, that he does not receive the penal reduction he would have had, had the proposal been accepted. Thus, on neither account does it seem that proposed neurointerventions involve exploitation.[16]

The previous considerations indicate why, more generally, I tend to believe that proposed neurointerventions are not properly characterized in terms of wrongful exploitation. It is very unclear whether an offender who is offered a pharmacological treatment or even some sort of invasive treatment in return for penal reductions can be said (or needs) to be either harmed or benefited

[15] Goodin uses a game analogy to illustrate what he means by playing for an unfair advantage. 'Fair play' is play according to the formal rules and informal ethics of a game. Taking unfair advantage, therefore, consists in availing oneself of 'strategic opportunities which are denied to one under the rules and ethos of the game at hand' (Goodin 1987, p. 183).

[16] This brief outline of various accounts of exploitation does not exhaust the range of theories presented. Some theorists have defended what could be referred to roughly as 'Kantian' accounts. For instance, it has been held that exploitation consists in treating others as a means only. As is well known, it is of course disputed what precisely this means. And it is far from clear that such an account is relevant in the present context if it is accepted that an offender can validly consent to a proposed neurointervention (see also Valdman 2009, p. 5). Another account has been to hold that exploitation involves degradation. We will return to a more thorough discussion of this objection in relation to neurointerventions in Chapter 4.

to an unfair degree (e.g. that the state benefits excessively). Furthermore, the basic idea across different accounts, namely, that exploitation involves taking advantage of a party and the unfortunate situation he or she is placed in, does not appropriately represent what is going on when treatment is offered to offenders. If there are sufficient moral reasons in favour of offering treatment (e.g. a crime-preventive effect) and if this is what motivates the offer, then the state is not simply taking advantage of the offender and his plight. The antidote-offering hiker is clearly trying to take advantage. But the state is not in a similar way trying to opportunistically profit from the situation. Rather, is it attempting to do what is morally required.[17] Needless to say, this is not tantamount to holding that a state cannot exploit inmates. If a local sheriff offers better prison conditions to those inmates who will help in the cleaning his private house, then this may well amount to some sort of exploitation. (It would constitute what some theorists regard as essential to exploitation, namely, an abuse of power). Or, if the state seeks to make big business out of using inmates as a cheap labour force, then the talk of exploitation may be apposite (see e.g. Browne 2007). However, surely this is something very different than, for moral reasons, the pursuing of a legitimate goal such as crime prevention. Finally, it is worth noting that several theorists agree that, even if exploitation constitutes a wrong, this does not imply that—everything considered—an exploitative act is always wrong (e.g. Feinberg 1983; Goodin 1987). However, while there are no countervailing moral reasons involved in standard examples of exploitation (such as in the antidote case), there will be strong countervailing reasons involved when we are considering crime prevention (e.g. the well-being and lives of potential victims). Thus, those who would object to proposed neurointerventions by insisting that offers of this kind to offenders are exploitative would still have to establish that such proposals are wrong on balance. Therefore, all in all, the reference to exploitation does not seem to provide a firm ground for rejecting the offer of combined neurointerventions and penal reductions to offenders.

The second objection to which we shall now turn is, from the outset, more promising. While considerations of exploitation, as indicated, usually

[17] It might perhaps be objected that, even if the state is trying to do what is morally required, this does not exclude the possibility that the offender is being exploited. For instance, it could be held that if the hiker with the antidote offers this to the bitten hiker—when the $20,000 can be used to save starving children—this may still involve an exploitation of the bitten hiker. However, even if this were to be true, it is much less obvious that this would also constitute an instance of *wrongful* exploitation. (And I tend to believe that, even if this were the case, it has to do with the excessive nature of the offer, which is—or at least need not be—paralleled in the case of proposed neurointerventions.)

concern offers presented in other types of context, the objection that has been presented by Lene Bomann-Larsen deals specifically with the proposal of neurorehabilitation for offenders. What she suggests is not that proposed neurointerventions in general are morally unacceptable. In fact, she presents her thoughts as offering a defence for the use of neurorehabilitation. But she argues that the use of such interventions should be narrowly constrained. More precisely, she contends that neurorehabilitative treatment 'should not go beyond what is necessary in order to correct the behaviour for which the criminal is imprisoned' (Bomann-Larsen 2013, p. 11). This conclusion is reached on the ground of a theory of morally appropriate and inappropriate offers.

In contrast to the previously considered standard objection, Bomann-Larsen does not believe that proposed neurointerventions undermine the possibility of obtaining consent. As long as an offender can choose between a fixed sentence and treatment accompanied by a penal reduction, she believes that consent can be 'effective', in the sense that it can be freely given. However, her point is that there are some offers which in themselves constitute a way of wronging the recipient and, when this is the case, the recipient's consent cannot take the wrongness out of the act of making the offer. But when does an offer constitute a way of wronging a recipient?

The answer is that this depends on the relation between the offerer and the recipient. For instance, there may be some offers which it is acceptable for a wife to make to her husband but which would be inappropriate if made between other parties who do not stand in the same personal relation. However, Bomann-Larsen also believes that there are some offers that are always inappropriate. In other words, what she holds is that 'for an offer to be appropriate it must be made by someone who is in the right normative position vis-á-vis the other, and some offers will be inappropriate *tout court* because no one is in the right normative position to make them' (Bomann-Larsen 2013, p. 74). But what then does this imply for offers made by the state to criminal offenders? At this point, Bomann-Larsen's suggests that 'what citizens are answerable for to the state determines the scope of behavioural conditions for which the state can appropriately offer convicts treatment; otherwise, the state illegitimately trespasses the borders of its jurisdiction vis-à-vis citizens' (2013, p. 74). This means that the state can only appropriately offer neurotreatment of the behaviour for which an agent is convicted in a particular case. If a treatment offer is not thus narrowly targeted it will be morally inappropriate. Bomann-Larsen's suggestion is interesting and apparently has gained some adherents (see e.g. McMillan 2014b). However, it opens two questions: first, should we accept this idea of some offers being wrong in themselves and, second, has the narrow domain of appropriate state offers been plausibly specified? In my view, there are reasons to be sceptical about both points.

Consider first the view that the speech-act of making an offer can in itself be morally inappropriate. An obvious example is: 'I will pay for medical help for your (otherwise dying) child if I get to have sex with you'. Bomann-Larsen holds that this constitutes an example of an offer which is always wrong in itself. No one can stand in the right moral position to make this offer appropriate. Other offers, as mentioned, may be inappropriate in some context but not in others. But why are some offers wrong in themselves? At this point, Bomann-Larsen has little to offer. She says that some offers constitute a violation of the other's 'claim to moral respect' and that the other is not being recognized as a 'moral equal' (2013, p. 9). In my view, these arguments give very little guidance with regard to distinguishing appropriate and inappropriate offers (see Ryberg and Petersen 2013). Moreover, it is not clear that such arguments provide sufficient support for the claim concerning the wrongness of particular offers. Suppose, for instance, that the offer of medical assistance in return for sex is presented in a very special context, namely, that it is the case that the offer-giver's only motive is to assist the dying child, that the only ways he can provide money to do so is by asking for sex with the parent (because some other person has offered money to someone who is willing to make the offer) and, finally, that for some reason it is not possible for the offer-giver to explain this to the parent. Admittedly, this context is highly hypothetical (which is why this kind of offer in real life would probably always be wrong). But it may nevertheless help us when it comes to the determination of the wrongness of offers. If, as suggested, there is no other way in which the offerer could possibly assist the dying child than by making the above offer, then this does not strike me as morally inappropriate. To insist that it is morally preferable not to engage in the mere speech-act of making the offer than to let the child die seems to me highly dubious. (In fact, I tend to believe it would under these circumstances be wrong *not* to make the offer and thereby let the child die.) Thus, in my view, there is reason to be sceptical about Bomann-Larsen's claim that there are some offers which are always wrong because no one is in the right moral position to make them.

However, be that as it may, suppose that we *arguendo* assume that Bomann-Larsen is right in holding that there are some offers which it is wrong to make. This leaves us with the second question, namely, why appropriate offers must concern only treatment that will prevent the precise type of crime for which an offender is convicted. I must admit that I fail to see why such offers must be this narrowly restricted. First, it is not absolutely clear to me what the restriction implies. For instance, suppose that an offender is suffering from a severe lack of impulse control and has been convicted for a particular crime C_1. Suppose further—as is usually the case—that a lack of impulse control can result in a set of future crimes C_1 to C_3. Does the restriction imply that it would be inappropriate, and therefore wrong, to offer a neurointervention that will prevent C_1 if this treatment will also prevent the offender from committing C_2 and C_3? In that case the implication of the restriction seems somewhat odd in light of the fact that this extra effect could, in some cases, be regarded as highly desirable by all parties including the offender himself. Moreover, there will probably in practice be very few cases in which one knows that a treatment will only prevent one particular crime. Alternatively, is the idea that the restriction in this case does not preclude the offered neurointervention as long as the treatment will prevent C_1? If this is Bomann-Larsen's idea than it seems that her position opens up for a very unrestricted view on appropriate offers. In fact, her position would be consistent with the offer of a treatment that will prevent the offender from committing all possible types of crime (because C_1 will be included in this set). Thus, it is far from clear precisely how the restriction should be interpreted.

Second, and more importantly, it is unclear why we should at all accept that an offer is appropriate only if it targets the particular crime for which an offender has been convicted. To claim that the criminal justice system is not concerned with crime prevention is not correct.[18] Obviously, crime prevention plays a different role in different penal theories. But all theories seem to leave some room for considerations of crime prevention (in fact, this is the case even for positive retributivist theories).[19] Furthermore, the idea that the state can only appropriately propose treatment to prevent C_1 (the crime for which the offender has been convicted) but not to prevent another crime C_2, finds very little support in the reasons Bomann-Larsen gives for why some offers are inappropriate. Suppose *ex hypothesis* that an offender

[18] For arguments to this effect, see e.g. Husak (2011). What he concludes is that: 'Surely preventing future harms is a legitimate function of the state—even through its institutions of criminal justice' (Husak 2012, p. 1203).

[19] For instance—and we will return to this (Chap. 4)—positive retributivists may well hold that crime prevention should play a role when it comes to the choice between different types of punishment with the same penal bite (see e.g. von Hirsch, Wasik, and Green 1989).

has been convicted of rape but that there are reasons to believe that there is a high risk that this person will commit assault in the future. To contend that the offer of a reduced time of incarceration in return for participation in a treatment reducing the risk of rape would be permissible, while a corresponding penal reduction in return for participation in treatment reducing the risk of assault would be inappropriate, because this would violate the offender's 'claim to moral respect' or fail to recognize the offender as a 'moral equal,' seems to me conspicuously unpersuasive. (Recall that, on Bomann-Larsen's view, the offender is free to accept or refuse the offer in both cases.)[20] Thus, the main problem in Bomann-Larsen's otherwise interesting analysis is that it fails to provide the requisite theoretical support, both when it comes to the view of some offers being wrong per se and, not least, with regard to the more precise restrictions that are held to follow from the suggested appropriateness-constraint.

To sum up, it may well be held that there can exist offers which are morally problematic even if they are not coercive and, therefore, do not deprive the recipient of the possibility of giving a free consent. However, as we have seen, so-called exploitative offers did not seem relevant in relation to proposed neurointerventions. Moreover, Bomann-Larsen's theory of inappropriate offers did not seem to provide a firm ground for placing narrow restrictions on the proper proposal of such interventions.

Penal Reductions and Penal Theory

Discussion in the previous sections reflects the way in which the idea of offering penal reductions in return for treatment participation has typically been considered. Adherents of such offers have emphasized the beneficial implications which treatment could have for the offender himself and not least for potential victims, while critics have underlined that such offers per se are morally unacceptable, either because they are coercive or perhaps for other reasons. Given the important role, as mentioned, which consent plays

[20] As another example, suppose *ex hypothesis* that an offender has been convicted of attempted murder and that there is a risk that she will commit murder after release. In this case, it seems clearly absurd to contend that, since the offender has been convicted of an attempt to murder, not murder itself, it would be appropriate to offer a neurointervention that would reduce the risk of future attempted murder but inappropriate to offer treatment that would reduce the risk of murder. Moreover, to hold that the former offer would treat the offender as an equal while the latter would not, seems hard to accept.

in modern medical ethics, this focus on whether or not a treatment offer is acceptable is not surprising. However, more surprising is the fact that the discussion has usually been limited to this narrow framework. Since the offer concerns the possibility of receiving a more lenient punishment, it is rather obvious that there is also a penal theoretical side to the discussion. As we shall now see, once we start to analyse proposed neurointerventions from a penal theoretical perspective, things become more complicated.

One of the things that complicates the discussion is that there is no general consensus on what constitutes the most plausible moral theory of punishment. Suppose first, therefore, that a plausible justification of punishment is provided by a consequentialist theory (e.g. a utilitarian approach). That is, that the infliction of a punishment on an offender is justified primarily as an instrument of crime prevention (attained, for instance, through deterrence, rehabilitation, or incapacitation of the offender). Suppose, furthermore, that the combination of the reduced punishment and a neurointervention is morally preferable from a consequentialist point of view.[21] In this case, it gets difficult to believe that the view, that proposed neurointerventions should be rejected because they are coercive, can plausibly be maintained. Strictly speaking, it is not inconsistent in itself to subscribe to a consequentialist justification of punishment while at the same time defending a constraint against coercive offers. But it becomes very hard to believe that it is possible to provide a coherent justification of why coercion is morally impermissible while at the same time subscribing to a theory that prescribes the use of compulsory punitive treatment of offenders for crime preventive reasons (see also Ryberg 2015b). That is, adherence to a consequentialist approach to punishment may well end up undermining the idea of coerciveness-based objections to proposed neurointerventions.[22]

Suppose, alternatively, that the most plausible approach to the justification of punishment is provided by retributive theory. How will this approach theoretically influence the analysis of offers of penal reductions? The answer

[21] It might be objected that, if the combination of a neurointervention and a reduced punishment is preferable to a more severe punishment from a consequentialist point of view, then there is no reason to *offer* this to the criminal. Rather, on a consequentialist account, this combination should simply be mandatory. However, this need not be the case. For instance, it could be the case that the neurotreatment would work well only if the offender accepts the treatment, but not if the treatment is forced upon him.

[22] To avoid misunderstandings, what I am claiming is not that a utilitarian theory of punishment commits one to utilitarianism in all other areas of ethics. The point here simply is that it is hard to believe that one can consistently defend the existence of a deontological constraint against coercion while at the same time accepting the use of compulsion for crime preventive reasons.

is that this depends upon the more precise content of the theory. If one subscribes to a version of *negative* retributivism, according to which there are upper proportionality constraints on punishment but where the proper punishment should be determined on the ground of consequentialist considerations below the upper constraints, then the idea of offering a more lenient punishment to offenders who accept a neurointervention may be fully acceptable.[23] However, this is no longer the case if one favours *positive* retributivism, according to which an offender should receive the precise proportionate punishment—that is, neither one that is more severe nor more lenient than that which the offender deserves. Following this view, it would be wrong to give the penal reduction to the offender who accepts the proposed neurointervention. If the offender deserves two years in prison but receives only one year accompanied by a neurointervention then this will violate the proportionality constraint. Thus, in this view it would be wrong to give an offender the offered penal reduction; and this is so, independently of whether or not the offer *in itself* is morally unacceptable (e.g. coercive or inappropriate).

In fact, matters are even more complicated. Previous considerations are based on the assumption that offenders are being punished as they ought to be punished. However, suppose this is not the case. That is, suppose that offenders in actual penal practice are being punished in a way that is too severe relative to their crime. This is precisely what many modern penal theorists believe to be the case, and we shall return to this later (Chap. 7). In this situation, the picture changes significantly. If it is (for political reasons) not possible to modify penal levels in accordance with what is ideally desirable then, under these conditions, it may well be highly desirable to offer penal reductions in return for participation in treatment programmes. This could be a way of reaching (or getting close to) the just punishment, that is, the punishment which the offender actually deserves on the ground of the severity of the crime. Under these circumstances, proposed neurointerventions could be a method of countering real-life over-punishment of offenders (Ryberg 2018). Thus, if the offer of neurointervention in return for penal reduction is not in itself coercive (or is unacceptable for other reasons), then there are

[23] The same is the case if one subscribes to so-called limiting retributivist theories, according to which there is, for crimes of varying gravity, a range of deserved punishment within which the proper punishment should be determined on consequentialist grounds (see e.g. Morris 1998; or Morris and Tonry 1990). According to these theories, one will just have to make sure that a penal reduction does not fall below the lower proportionality limits.

strong penal theoretic reasons in favour of making such offers. If, alternatively, such an offer is morally questionable in itself (e.g. because it is coercive) then one will have to consider how this reason against making the offer should be balanced against the penal theoretical reasons in favour of reaching a more just punishment (if the offer is accepted). In other words, under these circumstances it is not sufficient as an objection against proposed neurointerventions merely to hold that such offers are unacceptably coercive; one will also have to show that this constraint is morally more significant than the possibility of reducing disproportionate punishment. It is not obvious how an argument to this effect might sound like.

What these considerations clearly establish is that, in contrast to the impression one is left with in much of the modern discussion of 'coercive treatment', the question as to whether it is acceptable to propose neurointerventions to offenders as an alternative to reduced punishments, cannot be answered merely on the ground of a moral analysis of the nature of such offers (see Ryberg 2015b). One will (unsurprisingly, it might be added) also have to draw on penal theoretical considerations. And the more precise penal theoretical implications depend upon whether one subscribes to one theory of punishment or another, and upon what constitutes the actual circumstances under which such treatment proposals are presented.

Conclusion

Where do the considerations of the previous sections lead us? The discussion in this chapter has not so much been concerned with the various types of neurointerventions and with the question of how such interventions may differ morally (a question to which we shall return shortly). Rather, the point here has been to consider a first way of applying neurorehabilitation to criminal offenders, namely, by offering such a treatment scheme as an alternative to a longer but fixed period of incarceration. What we have seen, firstly, is that an objection to proposed neurointerventions based on the coercive nature of the choice facing the offender, had little strength. It was far from obvious that free consent constitutes a necessary requirement within a criminal-justice context and, even if this were the case, it is not clear that a choice between, on the one hand, a full fixed punishment and, on the other, a neurointervention and a penal reduction, undermines the possibility of obtaining a valid consent. On some accounts of coercion, this kind of choice did not constitute any threat to consent at all. (Or the proposal of neurorehabilitation could be presented in a way that would reduce the coercive force of the choice.) On other accounts, the choice would be non-coercive only under the assumption that the punishment given to the offender, if he does not accept the treatment-alternative, is justified.[24] Furthermore, we have seen that the idea that proposed neurointerventions should be rejected as constituting either exploitative or inappropriate offers did not seem convincing. Finally, it was argued that there is an often-ignored important penal theoretical side to the discussion according to which, depending on the precise content of the penal theory and the existing penal practice, there could be reasons both against and in favour of proposed neurointerventions.

In sum, I believe for the present we are left with the conclusion that, if there are strong moral reasons in favour of proposing neurointerventions to offenders (in terms of crime preventions, reduced costs, etc.) then: 1) it does not seem plausible to hold that such proposals should be rejected for reasons that have to do with the unacceptability of the offer per se; but 2) that there could, from some penal theoretical perspectives (such as, if one subscribes to positive retributivism) be reasons against such proposals. As we shall see in Chapter 4, however, this penal theoretical objection to proposed neurointerventions may be contested.

[24] As we have seen, even if this is the case, there could still be reasons in favour of proposing such interventions, namely, if this would lead to a punishment closer to the one which is morally desirable.

3

Compulsory Neurointerventions on Offenders

The discussion of the use of penal reductions in return for participation in treatment programmes involving neurointerventions, as we have seen, calls for the scrutiny of several ethical issues on which the views of theorists have been divided. However, in comparison, the use of neurointerventions as part of some sort of proposed treatment scheme is obviously much less controversial than the idea of using such interventions as part of an outright compulsory procedure imposed on an individual in order to reduce his or her propensity for crime.

A famous fictional example of a treatment designed to reform the mind of an offender—and one that has received renewed attention in the modern discussion of neurointerventions—is presented in Anthony Burgess's novel, *A Clockwork Orange*. In order to change his ultraviolent behaviour, the main character, Alex, is forced to keep his eyes open in order to observe on film the sort of brutality which he has committed himself, while, at the same time, a nausea-, paralysis-, and fear-inducing drug is injected into his bloodstream. The purpose of the therapy is to create an aversion to violent behaviour by causing a feeling of severe nausea when experiencing or even thinking about violence. Most readers of Burgess will be struck by the abhorrent character of this treatment and the novel has been interpreted as raising, in a compelling way, the question as to what constitutes an acceptable method of treating even someone convicted of brutal crimes. As Rodger Beehler has noted in his discussion of this sort of aversion therapy, many commentators have held that the novel depicts a picture of 'a method of prevention which . . . morality requires us to condemn' (1982, p. 647).[1] Noteworthily, he also suggests that if we were to consider the performance of a 'surgical operation on the brain' that would cause the same effect: many would probably be inclined to regard

[1] Beehler also notes that Burgess held that part of his purpose in writing the novel was to resist such methods of control (Beehler 1982, p. 648).

Neurointerventions, Crime, and Punishment. Jesper Ryberg, Oxford University Press (2020). © Oxford University Press.
DOI: 10.1093/oso/9780190846428.001.0001

this method as even 'more radical than the Clockwork Orange technique' (1982, p. 652). Beehler's point, however, was to question in several ways such an immediate inclination. The purpose of this chapter is to follow in the footsteps of Beehler's early discussion of such procedures, by considering the ethical legitimacy of the compulsory use of neurointerventions on offenders. Unsurprisingly, this issue has prompted numerous reactions. Before turning to the arguments, a few preparatory comments should be made concerning the framework of the discussion.

The experience of being forced to undergo a certain type of treatment is usually highly unpleasant, both for the individual upon whom the treatment is imposed and for other parties (not least his or her relatives). This is of course no less the case if the treatment involves some sort of neurointervention. Obviously, there will be great variations between the experiences of different types of neurointervention. It is moderately unpleasant to be forced to take a pill—such as a nutritional supplement—which has a modest and transient effect, but horrifying and traumatic to be forced to undergo a mental change through a surgical brain procedure. In between these polar extremes the experience will probably vary with the degree of invasiveness and the effects of a treatment. However, given the fact that an individual would very much prefer not to be the subject of forced treatment, the starting point of the discussion must be that compulsory neurointerventions are *pro tanto* wrong.

In this respect, the point of departure is precisely the same as the one that characterizes the ethical discussion of the legitimacy of state punishment. Placing citizens behind bars or depriving them of their property, when viewed in isolation, is morally wrong and, therefore, requires a persuasive justification. Thus, it is not surprising that some theorists have emphasized the ethical affinity between the compulsory use of neurointerventions and punishment. For instance, as Greely has noted: 'we send convicted criminals to prison without giving them another choice. Prison has enormous effects on their liberty, physical safety, and mental health. If we can do that why can we not require a direct brain intervention?' (Greely 2012, p. 164). And Douglas, along the same lines, asks 'what sets medical interventions apart from minimal incarceration and the other kinds of intervention to which criminal offenders become liable?' (Douglas 2014, p. 105). This parallel between punishment and compulsory neurointerventions is instructive. For instance, if one believes, as most theorists do, that the punishment of offenders can be justified, then one cannot at the same time consistently reject forced neurointerventions merely by contending that no one should be subjected to compulsion. Thus, in the following, I will not enter into a broader discussion of whether compulsion per se is morally wrong. Correspondingly, as we will see, there may be other reasons against the use of neurointerventions that cannot be held while, at the same time, accepting punishment as a morally required reaction to crime.[2] On the other hand, it is obvious that an inference from the fact that punishment can be morally justified to the conclusion that compulsory neurointerventions are justified, would be conspicuously premature. First, it could be the case that a punishment serves purposes that are not served by the imposition of neurointerventions. (We will return to this claim in the next chapter.) Second, it may be the case that compulsory neurointerventions involve a kind of wrong that is not involved in the infliction of punishment. It is this question that will be considered in the present chapter.

Another delimitation of the discussion concerns the fact that a forced neurointervention can in different ways be related to the punishment which an offender receives. For instance, one possibility is that a treatment is imposed upon an offender independently of the punishment. That is, if the proper punishment for a crime is two years in prison, then the offender will have to serve this sentence irrespective of the fact that he is also forced to undergo a neurointervention. Another possibility is to terminate the prison term after the treatment has been completed (as is the case for Alex in *A Clockwork Orange*). A further possibility is to give the offender who is undergoing a treatment a more lenient punishment than the one that would otherwise be warranted on the grounds of the crime. In the discussion of how the criminal justice system should deal with dangerous offenders, it has sometimes been suggested that detaining high-risk offenders in order to prevent future crimes can be justified only if the offender is properly compensated (see e.g. Schoeman 1981). Along the same lines, it could perhaps be held that a penal reduction should be given as some sort of compensation for the compulsory treatment. However, which of these models is the most plausible ultimately depends upon penal theoretical considerations and will not be discussed in the following.

[2] In fact, the parallel between punishment and forced neurointerventions may be instructive even for some abolitionists. For instance, if one defends a version of abolitionism based on the combination of consequentialism (or some mixed penal theories) and the empirical belief that punishment does not prevent more harm than it causes, then one would in theory accept punishment, had the empirical circumstances been different. Thus, it would be relevant, even for this type of abolitionist, to consider the consistency of views on forced neurointerventions and the in-principle acceptance of punishment.

Another important point to underline is, of course, that whether a certain type of neurointervention can be justified as part of a mandatory treatment scheme depends upon whether there exist other and preferable methods for reaching the same results. For instance, the application of deep brain stimulation, even if this per se is not unacceptable as a forced treatment, may be unjustified if there exist other less intrusive treatment options or if the same results can be reached in other ways, such as through traditional methods of resocialization. Whether there exist alternatives to forced neurointerventions, which in this sense are preferable, is of course an empirically contingent question always important to keep in mind, but this will not be considered in this chapter (see Chap. 7).

Finally, it should be mentioned that the purpose here is not to evaluate existing mandatory schemes involving compulsory treatment. There currently exist numerous types of such programmes—e.g. involving drug-dependence treatment—in many countries (programmes which have often been heavily criticized for being inefficient and for serious human-rights violations) (see e.g. Lunze et al. 2016). However, the present purpose is to focus only on some of the principled objections to the compulsory use of neurointervention as an instrument in crime prevention.

With these preliminaries settled, we can now turn to what *will* be discussed. The chapter proceeds as follows. In the first section, it will be considered whether some of the worries and objections that have been presented in the broader debate of the moral enhancement of humans can also be held to be warranted in our more limited discussion of neurointerventions on offenders. Subsequently, we will address the question whether forced neurointerventions should be regarded as a violation of an offender's right to physical or bodily integrity. Finally, we will turn to what seems to constitute the most prominent objection, namely, that forced neurointerventions violate the right to mental integrity or freedom of the mind. It will be argued that, on closer scrutiny, these objections are far less compelling than theorists have suggested.

Objections to Moral Enhancement

Though, as we have initially seen, the idea of changing the minds of criminals can be traced back to the early days of penological thinking, the modern discussion of the enhancement of humans by biomedical means has not had its main focus on criminal justice issues. For instance, Savulescu and Persson have defended the use of different types of enhancement, including moral enhancement—that is, roughly, the deliberate modification of people's dispositions and behaviour in order to make them more moral[3]—as a way of tackling the most serious global threats, or the risk of what they call 'ultimate harm'. The background of the recommendation is the diagnosed mismatch between, on the one hand, the 'myopic' nature of human moral psychology, that is, the fact that we are mainly concerned about people in close proximity or in the immediate future and, on the other, the development of technology that has enabled the exercise of influence all over the world and far into the future (Savulescu and Persson 2012). The alleged remedy to the threats provided by the progress of modern technology is to counterbalance it with the moral enhancement of humans. This ultimately led the authors to suggest that: 'If safe moral enhancements are ever developed, there are strong reasons to believe that their use should be obligatory, like education or fluoride in the water, since those who should take them are least likely to be inclined to use them. That is, safe effective moral enhancement would be compulsory' (Savulescu and Persson 2008, p. 174). Unsurprisingly, this proposal has provoked numerous reactions.

Several of the objections that have been advanced in the debate deal with the fact that, in order for moral bioenhancement to constitute a genuine remedy to global challenges, it would have to be applied society-wide. For instance, it has been held that such a universal application might well involve a threat to egalitarianism and other political ideals (e.g. Sparrow 2014b). Likewise, it has been suggested that this use of moral bioenhancement might undermine the existence of reasonable pluralism in a liberal society (Brooks 2012). Such objections lose force when enhancement is considered only in relation to a narrow section of citizens, such as, the subgroup of criminals on whom neurointerventions might constitute a possible treatment option. However, there are other arguments in the broader debate on moral bioenhancement that could also be relevant in the discussion of the use of compulsory neurointerventions on offenders. In the following, three such objections will be considered: first, that any morally acceptable attempt to change people's moral dispositions implausibly presupposes a yardstick for which dispositions are morally desirable; second, that moral enhancement

[3] There has been some disagreement on how precisely moral enhancement should be defined (see e.g. Beck 2015; Raus et al. 2014). However, in the present discussion, this controversy is not important.

in a particular sense deprives people of moral freedom; and third, that moral enhancement may affect the identity of individuals. However, as we shall now see, none of these objections provide sufficient grounds for rejecting the compulsory use of neurointerventions on criminal offenders.

The yardstick objection

The first argument to have received attention from several theorists roughly amounts to the claim that, in order for the idea of moral bioenhancement to make sense, one will have to know what 'moral' consists of. However, the answer to this question is far from simple. Sparrow has presented the objection in the following way: 'Moral questions are some of the most controversial matters in contemporary society and are fiercely disputed by different social groups. If we are going to start giving people drugs to make them more moral, we had better know what it is for someone to be more moral. But then who are advocates of moral enhancement to say that they know the answer to this question?' (Sparrow 2014b, p. 22). This overall concern can be given varying interpretations. Some theorists have underlined the lack of consensus on moral issues. Others have phrased the objection as a currently existing lack of knowledge of what is morally right and wrong. What this indicates is the fact that the objection can be given various interpretations in drawing on different views on the nature of morality (e.g. reflecting views on whether there are true—but yet unsettled—statements about moral facts; whether consensus or convergence constitute a truth mark; or whether moral facts exist at all). In the present context, it is not necessary to engage in depth in a consideration of how the objection might represent different meta-ethical positions. For our purpose, it will be sufficient to stay with the overall formulation that the project of moral enhancement is threatened by the lack of a proper yardstick for what would constitute genuine moral enhancement, that is, for what is right and wrong. The question is whether an objection along these lines can plausibly be held to have force in relation to our discussion of the use of neurointerventions.

Before answering this, it might be instructive to briefly return to the initial example of the attempt to reduce criminal inclinations through aversion therapy. Though Burgess' description of the Ludovico Technique may strike modern readers as a kind of treatment having little affinity with methods that criminal justice systems currently use in dealing with criminal offenders, it was certainly not a technique without resemblance to existing treatment schemes being used at the time the novel was written. Consider the following non-fictional procedures. A patient suffering from sexual deviation is given the nausea-causing drug apomorphine as part of a treatment carried out at two-hour intervals. As soon as the nausea occurred, a strong light would be shone onto nude photographs, the patient meanwhile being asked to engage in sexual fantasies. Every second hour during the nausea period, he would have to listen to tape recordings explaining his sexual preference, while attaching terms such as 'sickening' and 'nauseating' to the social consequences of this preference. The treatment would continue for 30 hours, and after 24 hours it would be repeated for another 32. The subsequent night the patient would be awakened every two hours in order to listen to explanations of the future positive consequences of living without his sexual preferences. This would be followed by three days of intense sexual therapy involving injections of testosterone (James 1962; Feldman 1966). Another example: a sexually deviant patient would be injected with apomorphine and shown sexually stimulating items immediately before nausea was induced. This treatment would be repeated every two hours, day and night. The patient would be kept awake in the night with the use of amphetamines. No food was allowed in the treatment period which would go on every night for one week. After an eight-day break, the treatment regime would be resumed (Raymond 1956; Feldman 1966).

These treatments constitute just two examples of real-life aversion therapy techniques. In an interesting (and appalling, one might add) article from 1966, Feldman provides an overview of a number of techniques applied in the 1950s and 1960s, often on patients referred by criminal courts. These techniques, as illustrated, typically involved combinations of nausea-causing drugs, injections of hormones, and electric shocks. The point of directing attention to such examples here is not simply to illustrate—as has often been the case in the treatment of criminal offenders—that truth is stronger than fiction, but rather to underline what the 'patients' were treated for. The regimens were imposed on homosexuals and transvestites.

With this in mind, it seems it is worth considering the yardstick objection. If it can be the case that some sort of behaviour is being criminalized at one point in time, only for it to become de-criminalized and accepted a few decades later, does this then threaten the idea of using compulsory interventions in the brains and minds of offenders? Applying the general yardstick objection to the present context, the conclusion that the criminal justice system should not adopt treatment schemes involving compulsory neurointerventions would be based on the premises that: 1) we do not possess a sufficiently certain yardstick of which sort of behaviour should be regarded as right or wrong; and 2) without a sufficiently certain yardstick, the criminal justice system should not adopt treatment schemes that involve compulsory use of neurointerventions. Is this argument convincing?

Despite its immediate appeal, I believe that the answer must be in the negative. Even if one accepts the claim that there exists widespread disagreement with regard to what moral enhancement in general would consist in, it is hardly plausible to maintain this claim with regard to many of the types of behaviour which are currently being criminalized. For instance, there exists a general consensus (also over time) that it is wrong to deprive others of their property or to physically assault innocent people. In fact, this has also been underlined by several exponents of the yardstick objection in the general discussion of moral enhancement. For instance, Sparrow notes that 'at least in some central cases, there would appear to be less controversy about what morality requires than advocates of cultural relativism or moral pluralism sometimes insist. Murder is wrong, rape is wrong, torture is wrong—and these truths are (almost) universally acknowledged' (Sparrow 2014b, p. 22). [4] In other words, a yardstick does seem to exist with regard to which crimes it would be morally desirable to prevent, which indicates that premise 1) must be regarded as dubious. [5]

However, it might perhaps be felt that this answer is too easy. After all, is it not the case that people in earlier periods (and in some countries today) would, with precisely the same deeply felt conviction, have insisted that homosexual practice constitutes an obvious example of a type of behaviour that is morally wrong and which ought to be criminalized? And does this not establish that we should not have too much confidence in our current yardstick for what is right and wrong? The problem with this claim is that, if one insists

[4] In the same vein, Nicholas Agar—another critic of Savulescu's and Persson's considerations on moral enhancement—contends that: 'A psychopath's lack of empathy provides a clear example of how a deficiency in a moral input can impair moral judgment. We can imagine a bio-medical moral therapy that morally improves a psychopath by restoring a normal aversion to inflicting suffering. Prison psychologists provide moral therapy to psychopaths by talking to them. There's no reason a drug might not have the same moral therapeutic effect' (Agar 2015, p. 343). For instructive considerations on a moral yardstick, see also Douglas (2008).

[5] An adherent of moral enhancement who makes a similar point is Degrazia who contends: 'Stick to improvements that represent points of overlapping consensus among competing, reasonable moral perspectives'. He outlines a list of what he believes all 'reasonable persons' could agree on that would represent 'moral defects' and, consequently, where there would be consensus on what moral improvements would consist in (Degrazia 2014, p. 364).

on subscribing to a strong interpretation of what it means to have a 'sufficiently certain yardstick', then premise 2) of the objection becomes dubious. The problem is that there are various ways in which we react to and try to prevent behaviour which we regard as wrong—and which few would believe should be abandoned even though we do not possess a sufficiently certain moral yardstick. For instance, most parents try hard to teach their children the difference between right and wrong. The same is the case in the educational system. Furthermore, as mentioned, most theorists today believe that the punishment of offenders can be morally justified. But if one insists on the view that taking proper action requires a sufficiently certain moral yardstick, then how could it possibly be justified to place some people behind bars for their misdeeds? Why should such a tangible procedure not require a sufficiently certain moral yardstick? It might perhaps be objected, that this way of questioning premise 2) is premature. More precisely, it might be held that the use of neurointerventions is different from various other (acceptable) ways in which people and society in general react to and seek to prevent wrongdoing: what one should not do without a proper yardstick is to cause changes that *permanently* affect people.

However, if this is basically what the yardstick objection amounts to, then it suffers from two obvious flaws. First, it is far from clear why the absence of a sufficiently certain moral yardstick should constitute a problem only for permanent interferences in people's lives but not for non-permanent interferences (e.g. it is hard to explain why placing an offender behind bars for 10 years does not presuppose such a yardstick, while a very modest but permanent change of a person's preferences or emotional reactions does). Second, and more importantly, the objection would succeed in ruling out only a very few types of neurointervention. Most sorts of neurointervention—such as pharmacological treatment and even some types of invasive techniques (e.g. deep brain stimulation)—last only as long as the treatment is provided (or only for a limited period after it is terminated). Moreover, though this admittedly may seem a more futurist scenario, it could be the case that even some forms of otherwise permanent changes could be reversed if, at a later stage, they were found undesirable. Thus, this interpretation of the objection is not persuasive.

Therefore, all in all, I believe that the yardstick objection should be rejected as relying on an exaggerated requirement of when interferences in people's lives are morally acceptable. Perhaps part of the immediate attraction, which admittedly is part of the objection, arises from the fact that it underlines that an acceptable use of compulsory neurointervention presupposes a plausible account of which types of behaviour should be criminalized at all or that there exists a risk that overzealous states might apply such mandatory treatment schemes in cases where this is highly unacceptable. However, the fear of the risk that states might misuse neurointervention is basically another argument than the yardstick objection that has been considered here (and one to which we will return in the final chapter).

The freedom objection

Another concern that has attracted considerable attention in the discussion of moral bioenhancement is that such enhancement methods, or at least certain variants of them, will restrict the freedom of enhanced individuals. This objection has most prominently been advanced by John Harris. Citing Milton's *Paradise Lost*, he contends that moral bioenhancement may undermine a person's 'freedom to fall' (Harris 2011, 2013a, 2013b, 2013c, 2014a). More precisely, what he has in mind is made clear when he explains that '[t]he freedom of which I (and I believe Milton) speak, the freedom to fall is the freedom to decide whether or not to fall for reasons, which have to do with what is best, 'all things considered'. Anything, which by influencing attitudes or emotional responses inhibits that ability significantly, is inimical to freedom' (Harris 2014a, p. 373). Thus, while he underlines the importance of traditional methods for moral enhancement such as 'bringing children up to know the difference between right and wrong, to avoid inflicting pain or suffering on or doing harm to others; and instilling in them habits or respect for others', the problem with moral bioenhancement is that it works as an 'incapacitator' destroying an understanding of good and bad and, more importantly, inhibiting the ability to do good over bad (2011, p. 104).[6] Without a freedom to fall, one cannot *choose* the good.

In considering whether this freedom-based concern can be transferred from the general discussion of moral enhancement to our present discussion of the use of neurointerventions on criminal offenders, it is worth noting that Harris' worry was anticipated by Burgess in *A Clockwork Orange*. Having gone through the treatment, the remarkable result of the conditioning

[6] As such, Harris believes that an enhancement that would only make us better agents but 'worse thinkers', cannot properly be regarded as moral enhancement at all (see e.g. Harris 2013c, p. 120).

technique is demonstrated to an audience of prison and government officials. Alex is humiliated but finds himself unable to engage in the violent response he would have chosen prior to the treatment (literarily, he ends up licking the boots of his aggressor). It is underlined that Alex has ceased to be a wrongdoer, but also that he has 'ceased to be a creature of moral choice' (Burgess 1972, p. 99). The underlying thought—expressed by the prison chaplain in the novel—seems to be that a man who can choose, but who chooses the bad, is in one respect better than someone 'who has the good imposed upon him' (see also Beehler 1982, p. 648). Does this concern make sense when a focus is placed on the use of neurointerventions on offenders in order to prevent future criminal conduct? That is, is it plausible to contend that mandatory treatment schemes involving such interventions are unacceptable because offenders will have their freedom compromised, not merely in the sense that the treatment is compulsory, but in the sense that the result of the treatment is a subsequent deprivation of their freedom to fall? In my view, there are several reasons to be sceptical about this objection.

First, let us *arguendo* assume that it is the case that a neurointervention will modulate an offender's mind in such a way as to be freedom-compromising in the outlined sense. The question that arises then is whether this suffices to establish that the treatment is morally unacceptable. The first thing that should be noted in this connection, as we have seen, is that if one accepts that incarceration can be used as an acceptable form of punishment of at least some offenders, then this is tantamount to accepting a considerable delimitation in the freedom of offenders.[7] An adherent of the freedom objection might of course object that what we are considering here is a different type of freedom restriction. While imprisonment places outer barriers on the freedom of the offender, a neurointervention could be held to place inner barriers in the mind of the offender. However, if this is the answer one resorts to in order to uphold consistency between the freedom-objection and the acceptance of imprisonment as a proper type of punishment, then it would be nice to know what constitutes the differences between these two ways of restraining freedom; that is, why the former is acceptable while the latter is not. A possible answer could be that it is unacceptable to interfere in people's minds. But in that case, the argument seems to collapse into an objection based on the right to mental non-interference that will be considered shortly. Another answer could be that the punishment serves a particular

[7] As Harris himself points out 'prison is normally considered the antithesis of freedom' (Harris 2014b, p. 249).

function that is not served by the treatment. However, this presupposes that neurointerventions cannot serve the function as a punishment. Whether this is a plausible claim is a question to which we return in the following chapter.

Furthermore, and more importantly, it should be kept in mind that what we are considering is the use of neurointervention as an instrument for crime prevention. But this means that if such an intervention succeeds in preventing the harm of future crimes—which of course is a morally highly desirable goal—then, in order for the freedom-objection to be conclusive, it would have to be shown that the way in which neurointerventions compromise freedom is either morally unacceptable—that is, that there is a moral constraint against this type of infringement—or that the value of maintaining freedom outweighs what is gained in terms of crime prevention. But it is far from clear why we should accept either of these views.

In many cases, we believe there exists a very strong reason to prevent individuals from being able to carry out criminal acts. And it is not even obvious that a neurointervention would have to *seriously* compromise an individual's freedom in order for it to prevent crimes. First, it seems that insofar as neurointerventions make it possible to narrowly target particular types of choice or act, there is no reason to believe that a crime-preventive intervention would generally undermine an offender's freedom to reflect on moral issues and 'to fall'. For instance, to stay with our literary example, even if it is the case that the Ludovico Technique has almost blocked Alex's ability to choose to commit a violent act, this does not mean that he has been equally deprived of the freedom to reflect on and choose other ways of acting. Degrazia has strongly underlined the same point by asking his readers to imagine a brain-implanted computer chip that robs an agent of one hundred per cent of his freedom with respect to rape and child molestation without diminishing his freedom to perform other types of action. In his view, 'the elimination of freedom with respect to these types of action is no great loss' (DeGrazia 2012, p. 367). Several other commentators have, in my view plausibly, made the same point (see e.g. Douglas 2014; Pugh 2017). But if it is possible in this way to more narrowly target particular types of choice and action, then this seems to indicate that the moral weight of a freedom restriction could be modest relative to the potential benefit in terms of crime prevention.

Second, it is not even clear that an offender need be fully deprived of his freedom to commit particular criminal acts. If the freedom to reflect on and choose a particular act is a matter of degree, then it could be the case that a minor reduction in an offender's freedom would in some cases be sufficient to cause a crime preventive effect. If this is so, then the costs in terms of a loss of freedom would also seem modest.[8] Thus, all in all, it is far from clear that what would be lost in terms of compromising the freedom to fall, constitutes a sufficient reason to regard this loss as morally unacceptable, everything considered.

However, perhaps this way of interpreting and meeting the freedom objection is too swift. It might be suggested—and perhaps this is basically what Harris has in mind in his criticism of moral bioenhancement—that the main point of the freedom objection is not to underline that freedom takes precedence over other values, such as the benefit of preventing future crimes, but rather to suggest that it is far from obvious that other types of benefit will in fact be accomplished if the method one is employing is to compromise people's freedom.[9] If freedom to fall, as Harris suggests, is the freedom to decide whether or not to 'fall for reasons', and if this freedom is lost or reduced by imposing a neurointervention that inhibits an individual's ability to perform certain acts—say, acts of violence—then the individual is also made incapable of choosing to 'stab or shoot someone who was trying to kill a loved one or an innocent person, if no less violent means of protection were available' (Harris 2014a, p. 373). As Harris points out, his 'concern is that there seems to be a real possibility that certain forms of MB [moral bioenhancement] will radically reduce the probability' of such ways of acting and that this may lead to 'disastrous results' (2014a, p. 373). In other words, and more generally, it could be held that the point of the freedom-objection is not to advocate the intrinsic or non-overridable value of freedom, but rather to suggest that if freedom is compromised by attempting to make people 'better doers' but 'worse thinkers' (Harris 2013c, p. 120), this may end up having deleterious consequences.

In this interpretation, the freedom-objection has clear affinity with another general worry that has been presented against the idea of moral bioenhancement, namely, a particular way of acting may be right in one context but wrong in another, and that a method that inhibits certain ways of acting may therefore not succeed in accomplishing what this method

8 Along the same lines, DeGrazia asks his readers to imagine that there exists a reasonable way of quantifying a reduction in our freedom and that, in comparison with the average degree of freedom, moral bioenhancement has reduced our freedom by twenty-five per cent but has also resulted in a world in which there are no wars and no starvation. Given this scenario, he 'would accept this reduction in freedom across the board if it were necessary for such great improvement in people's moral behaviour and such welcome consequences' (Degrazia 2012, p. 367).

9 If the objection in this interpretation were plausible, then it would not only constitute an argument against compulsory use of neurointerventions, but also against *proposed* neurointerventions as considered in the previous chapter.

is meant to accomplish.[10] Turning to the use of neurointerventions on offenders, it is clear that if it turns out that such interventions would not only inhibit future criminal activity but also the performance of various acts that are usually regarded as morally desirable, then this is something that should be taken into account in an everything-considered decision of whether such interventions should be effectuated. However, it is far from clear that neurointerventions designed to prevent future criminal activity would have such consequences. In particular, if it is possible to design such interventions to narrowly target certain types of acts, such as those sexually-motivated or violent, then it becomes hard to believe that they will eventually turn out to be counter-productive in the outlined way. Moreover, further support for this view is given by the next answer to the freedom-objection to which we will now turn.

The second reason to be sceptical about the significance of the freedom objection pertains to the question as to whether there are reasons to believe that crime preventive neurointerventions will actually compromise freedom in the outlined way. In the general discussion of moral enhancement, several commentators have argued that such enhancements need not compromise freedom (see e.g. Degrazia 2012; Earp et al. 2017). I believe the same can be held with regard to crime-preventive neurointerventions. The question is, how must a neurointervention influence an offender in order to prevent him or her from engaging in future criminal activity? Roughly, this can happen in three ways: 1) by influencing the offender's cognitive capacities; 2) by influencing motivational factors; or 3) through some sort of more direct influence on the offender's behaviour.[11]

The first type of influence can take different forms. As a simple example, suppose that an offender possesses only limited cognitive skills for moral reasoning. He may know that it is illegal to break into someone's flat to steal the property and that it will be unpleasant for the owner. However, he may not be able to fully comprehend the harmful consequences of his act. He may not grasp that he affronts the victim's sense of personal space, or that his entry is likely to cause a feeling of vulnerability or insecurity in the victim

[10] For instance, Sparrow mentions that '[e]ncouraging empathy and a sense of justice in individuals may usually be a good thing, but enhancing either of these faculties may make individuals more likely to behave unethically in various situations, as when, for instance, a judge declares a person "not guilty" as a result of empathic concern for them, despite the presence of overwhelming evidence that they are in fact guilty of the crime with which they are charged, or when a parent neglects a child out of an excessive concern for duties of justice toward strangers' (Sparrow 2014b, pp. 20–21).

[11] See also the discussion in DeGrazia (2012).

for a long period of time after the burglary. Suppose, furthermore, that a neurointervention will either directly contribute to improving the offender's limited capacity to fully grasp the harmful consequences of his potential future acts, or perhaps that it may indirectly provide a precondition for this type of improvement by opening the offender's mind to such reasoning and knowledge if it is sustained in other ways (e.g. through information or some sort of non-medical therapy). Would such an intervention compromise the offender's freedom? The answer is obviously in the negative. Surely, the offender's capacity to make decisions on the grounds of reason and, as Harris underlines, the ability to make 'all things considered' judgments, has not been reduced in any way by such an intervention. In fact, given the emphasis which Harris gives to the view that moral enhancement must, in large part, consist of cognitive enhancement, it seems that this type of neurointervention is one that he *ceteris paribus* should endorse (see Harris 2011, p. 106).

What then if we turn to the second possibility, that is, the idea of affecting an offender's criminal tendencies by using a neurointervention that will influence the motivational background for his future behaviour? Obviously, such influences can take various forms, and the purpose here is not to provide a general overview. However, to take a few examples: suppose that a neurointervention will reduce the strength of certain preferences, for instance, by reducing a certain sexual desire or by putting a damper on impulsive aggressive reactions. Or, alternatively, that it will increase the strength of certain preferences, such as a feeling of care or sympathy for others. Would a neurointervention with such effects compromise an offender's freedom? Once again, the answer seems to be in the negative. A change in the offender's preference profile may lead to a change in this person's behaviour—which of course would be the point of the intervention—but there is no reason to believe that such an intervention would compromise the offender's freedom to 'decide whether or not to fall for reasons' (Harris 2011, p. 373) or, in short, that it will turn the offender into what Harris calls a worse thinker (Harris 2013c, p. 120).[12]

[12] As mentioned earlier, Harris holds that anything that by 'influencing attitudes or emotional responses' inhibits the freedom to fall for reasons and to make 'all things considered' judgements is inimical to freedom. Though he also underlines that he is not suggesting that moral bioenhancement necessarily always has this effect, it is not very clear what type of enhancements would in fact have this effect. For instance, suppose an individual has been enhanced to possess a more caring or sympathetic attitude to other persons. Why should such changes in the attitude to others inhibit this individual's ability to reflect on and choose in accordance with reasons? It might well be the case, that the enhancement has caused a change in the weight that is attributed to some of the reasons, but surely this is not the same as holding that the individual has become less open to reasons or a 'worse thinker'.

But what then if we turn to the third possibility, namely, that a neurointervention may in a much more direct way influence an individual's behaviour? Would such an influence not compromise freedom by inhibiting an offender's ability to exercise reasoned decision-making? The question which this sort of influence raises is obviously what it means to influence an individual's behaviour in a way that cannot simply be reduced to changes in his or her motivation or reasoning. The most radical answer could be to point to situations in which an intervention results in certain bodily reactions. An interesting example is found in experimental studies of brain-computer-brain interfaces. For instance, in a recent study, a subject was trained, with the use of EGG-technology, to control the curser on a computer screen simply by focussing his mind on the curser's movements (Rao et al. 2014). This ability was used in a computer to determine when a cannon should shoot at enemy rockets. However, when the subject decided to shoot, the EEG-signal was transferred to another subject whose brain was simulated, by transcranial magnetic stimulation technique, to press his right hand on a touchpad, which caused the cannon to fire. In short, the first person decided when to shoot, the second did the shooting. Now, if this is how a neurointervention forced onto an offender works, that is, if it simply prompts a certain physical reaction in the offender, then it is certainly reasonable to hold that it would circumvent reasoned decision-making. The same could be the case if such interventions caused some sort of impulsive reactions. In fact, if I understand Harris correctly, this seems to be how he believes that different sorts of moral bioenhancement would work. As he underlines, such enhancements 'make it more probable that those who use them will act impulsively rather than thoughtfully' (2013c, p. 120). It is not absolutely clear to me why Harris believes this is how moral enhancements will (typically?) work. But even if we ignore this question, there seems to be no reason to hold that this is how crime-preventive neurointerventions will usually work. As indicated, such interventions might affect an offender motivationally without resulting in purely impulsive or automatized behaviour (which on standard accounts would in no way count as genuine acts). And even if this would be how some neurointerventions work, there would still be others not affecting a person's behaviour in this direct way—say, an intervention that would increase impulse control—which means that the freedom objection will be pertinent only in a limited number of interventions.

Then, where do these considerations lead us? What the discussion shows, in my view, is that the worry that Harris has formulated in relation to moral enhancement does not seem to constitute a genuine concern in relation to our present discussion of the use of neurointerventions as an instrument of crime prevention. First, it has been indicated that, insofar as such neurointerventions constitute a threat at all to the freedom of offenders, this will only be the case for a subset of interventions. In fact, it could even be the case that some interventions might improve an offender's ability to engage in reasoned considerations. For instance, insofar as an intervention will improve an offender's impulse control, or perhaps weaken the influence of certain overwhelmingly strong desires, this could increase his or her ability to reflect on and choose a course of conduct 'all things considered'. As mentioned in the previous chapter, this is precisely what has been suggested in relation to the treatment of certain sexual desires which are regarded as impediments to free or autonomous decision-making. Second, it has been argued that, even if, with regard to some neurointerventions, it is the case that they are inimical to freedom in the suggested sense, it is questionable whether this suffices to establish that the use of such interventions is morally unacceptable. Thus, in the present context, the outlined concern for freedom does not seem to constitute a convincing objection.

The identity objection

While the freedom objection, as we have just seen, has been presented narrowly in relation to the discussion of moral enhancement, there is another concern that has been advanced both in the discussion of *moral* enhancement and, more generally, in relation to the use of enhancement technology (see e.g. Bublitz and Merkel 2009; Degrazia 2000, 2005; Erler 2011). Broadly, the concern has been phrased in terms such as: biotechnological enhancement may 'turn the person into someone else', 'change who a person really is', or 'constitute a threat to the person's true self'.[13] For instance, Carl Elliott has held that: 'What is worrying about so-called 'enhancement technologies' may not be the prospect of improvement but the more basic fact of altering oneself, of changing capacities and

[13] Historically, it is precisely contentions like these that we were presented as early as in the mid-1940s when it was slowly being recognized that the widespread use of lobotomies was not without a price. For instance, Swayze quotes a wife to a lobotomized patient as saying: 'His soul appears to be destroyed; he is not the man I once knew'. The mother to another patient put it this way: 'She is my daughter but yet a different person. She is with me in body but her soul is in some way lost' (1995, p. 507). We will return to the dark history of neurointerventions in Chapter 6.

characteristics fundamental to one's identity . . . [D]eep questions seems to be at issue when we talk about changing a person's identity, the very core of what the person is' (Elliott 1999, p. 28). Along the same lines, the *President's Council on Bioethics* has warned that by becoming better than we are 'we risk 'turning into someone else', confounding the identity we have acquired' (2003, p. 300).[14] In order to examine whether concerns pertaining to personal identity constitute a problem in the present context, that is, in relation to the compulsory use of neurointerventions, it is necessary to clarify what the effects more precisely amount to, and what constitutes the moral significance of such effects.

While part of the discussion has arisen in relation to the use of neuropharmaceuticals, such as the treatment of patients with dopamine-sensitive drugs, the most recent focus has been placed on the consequences that deep brain stimulation (DBS) may have with regard to identity. Though DBS has been used as a treatment of patients with epilepsy and various types of psychiatric disorder, the main experiences stem from the comprehensive use of DBS in relation to patients suffering from Parkinson's disease. In such patients, electrodes are implanted to stimulate the subthalamic nucleus in order to alleviate movement disorders. While implants in many cases have highly beneficial effects on motor symptoms, they may also sometimes have serious psychological and psychiatric side-effects. As a commentator even said, the 'risk of becoming another person following surgery . . . is one of the most urgent conceptual and ethical problems facing deep brain stimulation in Parkinson's disease at this time' (Witt et al. 2013, p. 499). Possible consequences pertaining to identity have also, albeit to a minor extent, been considered in relation to other types of non-invasive or minimally invasive neurotechnology such as transcranial magnetic stimulation (TMS) and transcranial direct current stimulation (tDCS) (see e.g. Cabrera et al. 2014).

For a clearer picture of what this concern amounts to, it will be helpful first to distinguish between different kinds of identity-related consequences. The purpose here is not to suggest that the following categories present strictly separable and non-overlapping types of consequences, but rather to get an overview of some of the different consequences that may be relevant for an ethical assessment. Roughly, three overall categories of consequences can be outlined.

[14] Adina Roskies has, along the same lines, asked 'Will certain . . . therapies change who we are?' (Roskies 2007, p. 22).

The first type of possible consequence consists of cases in which an individual, literary speaking, becomes someone else. That is, more precisely, cases in which the *numerical* identity of the person is lost. As a hypothetical example, Craig considers what he calls the 'Damascus Drug', one that permanently changes a violent psychopath into a religious pilgrim (Craig 2016, p. 112).[15] The person is no longer one and the same. Precisely what kind of continuity is required for it to make sense that a person at one time is the same person at he is another time is disputed, and is not a question that we need to dwell on here.[16] What matters is that if a person has lost his or her identity in this sense, then he or she has ceased to exist.

The second type of consequence consists in changes of what could roughly be called the *personality* of an individual.[17] Changes in this category comprise alterations of a person's cognition, emotions, motives, and behaviour.[18] Some changes may concern core elements of who a person is, while others may be seen to be more peripheral. However, the important thing is that it still makes sense to hold that *the person* has undergone a change of personality; that is, that there is someone, numerically speaking, who has been changed. It is not difficult to find examples of how neurointerventions may cause changes that fall in this admittedly very broad category. For instance, in relation to the use of DBS on Parkinsonian patients, several severe changes in the traits of these individuals' personality have been reported (e.g. mania, hypersexuality, depression, pathological crying) (see e.g. Jotterand and Giordano 2011; Witt et al. 2013). Such changes will of course usually be recognized by the patients (as by their relatives), but this need not be the case.

The third category of consequences that might follow from the use of a neurointervention concerns the patient's self-perception or self-awareness; that is, roughly, how the individual experiences him- or herself after an intervention has been performed. For instance, some Parkinsonian patients have reported postoperative experiences of unfamiliarity with themselves or

[15] Even though Craig mentions that the Damascus Drug affects the person's 'personal identity', it is not clear whether he believes the effects to be as radical as to involve a change in *numerical* identity.

[16] For a classical modern discussion of personal identity, see Parfit (1984, part 3). See also Degrazia (2005).

[17] It should be underlined that I do not here use the term 'personality' in a precisely defined psychological sense (see e.g. the discussion in Witt et al. 2013).

[18] Historically, one of the most appalling examples is of course the various personality changes that were caused to patients who underwent a lobotomy (see Chap. 6). These patients were described by doctors and others as: 'dull, apathetic, listless, without drive or initiative, flat, lethargic, placid and unconcerned, childlike, docile, needing pushing, passive, lacking in spontaneity, without aim or purpose, preoccupied and dependent' (Swayze 1995, p. 508).

feelings of strangeness (Baylis 2013). Some patients complain that 'I have not found myself again after the operation' or 'I do not feel like myself anymore' (Baylis 2013, p. 514; Johansson et al. 2014, p. 30). Moreover, patient testimonials have reported, more narrowly, experiencing the existence of a foreign object in the brain and a stimulator implanted under the skin (usually just below the collarbone). For instance, patients have held that they 'feel like a machine' and even said 'I feel like 'Robocop' (Johansson et al. 2014; Witt et al. 2013). These experiences have often been described in terms of alienation, self-estrangement, or lack of authenticity (see e.g. Gilbert et al. 2017; Kraemer 2013; Pugh et al. 2017).

In outlining the three categories of consequences, it should of course be underlined that there may be major differences depending on what type of neurotechnology we are considering. Furthermore, several technical questions arise when such consequences follow from the use of a neurointervention. For instance, even if we focus only on the use of DBS, despite the fact that more than 100,000 patients worldwide have implants, only a relatively small number of studies on the postoperative changes of personality and self-perception in patients have been conducted. As Fredric Gilbert et al. have recently noted: '[w]ith so few phenomenological investigations into the experience of self, the philosophical debates about potential postoperative effects of DBS on the self are mostly based on anecdotal cases that may not be representative' (Gilbert et al. 2017, p. 97). Further knowledge of the extent to which the different sorts of consequences occur are of course important for an assessment of the *pros* and *cons* of a neurointervention.[19] Moreover, there are well-known problems when it comes to the comparison of the effects of the use of a certain technology between different groups of patients. For instance, in a recent study conducted on 17 Parkinsonian patients with DBS implants, the researchers found that 'the more patients preoperatively felt alienated by their illness, the more they experienced postoperative self-estrangement' (Gilbert et al. 2017, p. 98). Thus, there may be a relation between the specific disease that is being treated and the postoperative effects of the treatment. As has been pointed out, when DBS is used on Parkinsonian patients it affects a homeostatic system that has been modified by years of preceding neurodegeneration (Mosley et al. 2014, p. 54). But this means that one cannot take it for granted that the identity-related effects of

[19] Noteworthily, in their recent study, Gilbert at al. conclude that 'this work does not advocate that DBS alone directly causes potential personality changes' (Gilbert et al. 2017, p. 106).

DBS will be the same if the technology is administered to other groups of patients or, for instance, individuals suffering from aggressive impulsivity. Finally, it is clear that some effects can be avoided if a technology improves, and it is sometimes the case that effects can be alleviated when they occur (e.g. it has been reported that patients who struggle to 'find themselves' post-operatively often respond to a repletion of dopamine (Mosley et al. 2014)).

The point of these considerations is not to suggest that the identity-related effects of the use of DBS should not be taken seriously, but rather to underline that there are various reasons to be cautious when we consider the scope of the problem and, not least, the different ways in which the problem could be handled. However, this still leaves us with the basic question concerning the ethical significance of identity-related effects. If DBS or another type of neurointervention results in changes to the identity of a person, in a person's personality, or his or her self-awareness, then what does this imply with regard to the question of using such methods to prevent the future misconduct of offenders? The answer, I believe, is that this very much depends on the precise nature and severity of the effects.

The first possibility, that a neurointervention results in a genuine shift of numerical identity, is of course the most radical type of effect. It is the least likely category of effects—to my knowledge, none of the studies that have dealt with the question of neurointervention-induced changes of identity seem to be concerned with identity in this sense—and, in the present context, it is also the simplest question to handle. If a person, as a result of an intervention, ceases to exist, then this constitutes an effect of a degree of seriousness that reaches beyond what we are considering in this book. The purpose here is not to engage in the traditional controversy of ethical theory as to whether it can ever be justified to sacrifice the life of one person in order to save others.

What then if we turn instead to the two other categories, that is, the effects on personality and self-awareness? An interesting question in this relation is how would such effects affect the individual upon whom a neurointervention is imposed? In the discussion of DBS-induced changes of personality, it can sometimes sound as if all such changes are for the worse. Any change is described as a 'threat' to personality. But, obviously, this is not always the case. Baylis appears to be right when he rhetorically asks whether we 'would find the conclusion that DBS is threatening to personal identity as intuitively appealing' in cases where a depressed person has become happy or a person with obsessive compulsive disorder has been relieved of his impulses (Baylis

2013, p. 522). However, it is clear that there are changes in personality traits that are highly undesirable to the person him- or herself, as well as to others. The same is true in regard to changes in self-awareness. The claim that 'I do not feel like myself' will usually be an expression of frustration. Moreover, it might be held that any changes in personality traits or self-awareness—even if they are not per se for the worse—should be seen as morally dubious if they are forced upon an individual. However, to contend that there is a moral constraint against causing such effects in offenders seems to constitute too strong a position.[20] Without further qualification, it is hard to defend this contention if one at the same time subscribes to the view that the imprisonment of (some) offenders is morally acceptable (or required). Few (if any) theorists today seem to accept the outlook sometimes referred to as 'the deep freeze view' on imprisonment, namely, that an offender remains totally unaffected and invariant to periods of incarceration. On the contrary, several types of psychological effects have been suggested as following on from imprisonment. For instance, Adrian Ground has argued that some of the effects of imprisonment can be regarded as an 'enduring personality change' (see Liebling and Maruna 2005, p. 15).[21] And, as Alison Liebling and Shadd Maruna have emphasized in their comprehensive collective work on the effects of imprisonment: '[f]ear, anxiety, loneliness, trauma, depression, injustice, powerlessness . . . and uncertainty are all part of the experience of prison life' (2005, p. 3). If these empirical results are correct, then it is hard to reject the compulsory use of neurointerventions on the grounds of consequences concerning personality changes and frustrating experiences of lack of self-awareness, while at the same time adhering to the use of the incarceration of offenders.[22]

But if the identity objection is not sufficiently strong to reject the use of neurointerventions, then how should we regard the consequences that are outlined in the objection? The most plausible answer, in my view, is to consider the objection as directing attention to the serious sets of side-effects that might follow on from the use of such interventions. As such, the objection is

[20] More will be said on this shortly as part of the forthcoming discussion of the right to mental non-interference.

[21] Liebling and Manuna also mention that research on psychological effects has found, in certain groups of prisoners, symptoms such as 'persistent feelings of being "different" from others and from one's previous self' (Liebling and Manuna 2005, p. 15).

[22] Obviously, the fact that one accepts the use of imprisonment of offenders does not commit one to accepting all the types of incarceration that are currently being used (such as e.g. the conditions in maximum security prisons). However, even more decent ways of incarcerating offenders (as well as other types of contact with the criminal justice system) may have a negative impact on the offender, for instance, by causing experiences of frustration.

important as a reminder of the fact that, once one starts to interfere with the brains of individuals, this may (in some cases) have various sorts of side-effects beyond those that are perhaps associated more narrowly with medical treatment (e.g. the risk of infections when DBS batteries are shifted). But these effects do not morally prohibit the use of such interventions. Rather they should, if they occur, be balanced against the benefits that might follow for the offender himself, for relatives, and for potential crime victims.

The Right to Bodily Integrity

As mentioned, the objections that we have considered in the previous section have all been advanced as part of the discussion of the moral legitimacy of moral enhancement; that is, in a more general context. The same is the case for the objection to which we will now turn. However, the context is different and even broader. The idea that humans possess a right to bodily or physical integrity, that is, that our bodies are somehow morally protected, has been presented in relation to a wide range of moral issues, such as, male and female circumcision, the selling of bodily organs, transplantation, prostitution, the use of blood samples or other bodily extractions for scientific or other purposes—to mention but a few (see e.g. Dekkers et al. 2005; Ramachandran 2009). Moreover, it seems *prima facie* obvious that such a right could be relevant in relation to the question of the use of neurointerventions as part of mandatory treatment schemes. The non-consensual injection of a neuropharmaceutical or a neurosurgical procedure imposed upon an individual seems, quite obviously, to constitute an unwanted incursion into the individual's body and, hence, to violate this individual's right to bodily integrity. Thus, it is not surprising that some attention has been paid to this right. For instance, as part of his comprehensive criticism of the use of castration of sexual offenders in the US, John Stinneford contends that both surgical and chemical castration 'involve an invasion of bodily integrity' (Stinneford 2006, p. 596). Along the same lines, Olivia Choy et al. hold that 'forcing biological interventions violates an individual's right to bodily integrity' (Choy et al. 2018, p. 4). Other theorists have expressed similar views.[23] Put generally, the objection that compulsory neurointerventions are morally

[23] For instance, Richard G. Boire holds that 'compelling a person to use a pharmachotherapy drug . . . implicates the person's traditional rights to bodily integrity' (Boire 2010, p. 288).

unacceptable because they involve such a rights violation is based on the premises that: 1) compulsory neurointerventions intrude on an offender's bodily integrity, and 2) intrusions on someone's bodily integrity constitute right violations and, hence, are morally unacceptable. Let us consider each premise in turn.

In order to assess premise 1), the first thing needed is a clearer picture of what it means to intrude on someone's bodily integrity. Though, as noted, the concept has figured as an objection against various sorts of practice, it does not take much scrutiny to see that it is far from obvious what kind of interference the right to bodily integrity allegedly protects an individual against. Here are possible interpretations.

(A) The first possibility is to hold that bodily integrity concerns an individual's ability to move his or her body as he or she wishes. On this account, a person's right to bodily integrity has been infringed if the person's movements have been restrained by external causes (e.g. if the person's hands or legs have been tied) or if he or she has received a drug that makes it impossible to move his or her body, or parts of it.

(B) Another account has been to relate bodily integrity to the operations or basic functions of the whole body or its parts. This account has sometimes been considered under the term of 'functional wholeness' (Dekkers et al. 2005, p. 184). The idea is that individual body parts or the body as a unit is biologically designed to serve certain functions and that interference with such bodily functions constitutes a violation of a person's bodily integrity. However, in the absence of further qualifications, it is not clear what this precisely means. First of all, some sort of specification of what counts as a relevant bodily function is required. Moreover, it may be the case that the function of an individual body part can be affected without affecting the overall functionality of the body (e.g. the function of a kidney is affected if this particular organ is destroyed, but the overall bodily processes of filtering bodily fluids may still be preserved by the other kidney). And it also needs to be clarified whether any kind of effect caused to bodily functions should be regarded as an intrusion of bodily integrity, that is, for instance whether a mere change of a bodily function is in this respect sufficient, or whether this requires that certain functions are reduced or totally obstructed. However, for our present purposes, we need not engage in further explications of this account.

(C) A very different approach has been to relate bodily integrity to an individual's bodily feeling. More precisely, what matters is not simply the various types of feeling that may be caused when the body is affected but rather the more precise experience of what has been called 'bodily wholeness' (Dekkers et al. 2005, p. 184). Interestingly, by being a purely experiential approach, this account of bodily integrity does not imply any direct relation between the anatomical wholeness of a body and the individual's bodily integrity. Several (rare) disorders exist that imply a mismatch between an individual's physical or objective body and his or her experienced body. For instance, people suffering from Body Integrity Identity Disorder (BIID) may experience one or more limbs as an appendage to their body from which they strongly desire to be extricated.[24] In a number of cases, amputations of healthy limbs have been performed on patients suffering from BIID, both in order to remedy an intense and persistent feeling of alienation towards the limb and to prevent the patients from mutilating themselves (see e.g. Müller 2009). Another complementary phenomenon is the experience of phantom limbs. Whereas an individual suffering from BIID lacks the awareness of an existing limb, the experience of phantom limbs consists in the awareness of limbs that have been removed or are congenitally missing (e.g. Sobchack 2010). Thus, an intrusion on a person's bodily integrity in this sense involves some sort of influence on a person's bodily feeling and is not always in any simple way related to the person's material body.

(D) Another interpretation of the concept concerns the way in which the outer physical boundary of the body is affected by others. For instance, an obvious example of a violation of a person's bodily integrity in this sense would be a case in which a person has become the victim of unwanted sexual touching.

(E) Yet another account has to do with entrance through the outer physical border of the body. In this interpretation, a violation of the right to bodily integrity would involve cases in which the outer surface of the body is non-consensually penetrated.

(F) Finally, in a very broad interpretation, bodily integrity is intruded upon by any unwanted alterations in the body. In so far as it is possible to cause changes inside the body without penetrating its surface (e.g. by the use of a gamma-knife), such non-consensual changes would constitute violations of the right to integrity in this particular sense without threatening integrity in the previous sense. This account of bodily integrity has sometimes been presented in terms of bodily intactness (Dekkers et al. 2005).

[24] In some extreme cases, patients suffering from Pötzl Syndrome seem to ignore parts of their body or even to perceive a limb as part of another person's body (see Müller 2009, p. 38).

As this enumeration of possible interpretations shows, the concept of bodily integrity is far from being unambiguous. Obviously, this does not mean that a proponent of a right to bodily integrity will have to subscribe exclusively to one of the above interpretations. It might well be the case that the right comprises several of the accounts and thereby provides moral protection against several of the outlined types of intrusion on the body. However, the point is that the concept can be given different interpretations and that an explication of these interpretations is required in order to examine whether premise 1) of the objection is true; that is, whether compulsory neurointerventions can plausibly be held to involve an intrusion on bodily integrity. Suppose we distinguish between the following four ways in which a neurointervention can be imposed: as a pill; an injection; a surgical procedure (such as a DBS); or a non-invasive stimulation (e.g. transcranial magnetic stimulation (TMS)). The question then arises whether any of these methods constitute intrusions on bodily integrity in any of the outlined interpretations.

From the outset it is clear that several of the outlined interpretations are irrelevant given these four types of intervention. None of the accounts (A) to (D) seem relevant in the present context. Forcing someone to take a pill, having an injection, undergoing a surgical procedure, or a non-invasive stimulation of the brain, obviously does not place restraints on the movement of the body. Neither is it reasonable to hold that such interventions threaten any bodily functions.[25] Even though it is correct that some sort of neurological change will be caused by the interventions, this is hardly the same as holding that part of the brain's central functions have been blocked. And it does not seem that any of the four methods threaten bodily integrity by affecting the outer surface of the body. Even though an injection and a surgical procedure penetrate the skin, neither procedure seems to be an infringement on someone's bodily integrity in the same way as if the person were the victim of unwanted sexual touching. However, the picture changes if we turn to interpretations (E) and (F). An injection and a surgical procedure each constitute a way of penetrating the physical boundary of the body. The same, I guess, can be said to be the case if a person is forced to take a pill. Notably, however, it is less obvious that this can also be held to be what happens if an individual is forced to go through some sort of treatment involving TMS. What precisely

[25] Even though I do not believe that any of the four types of intervention can plausibly be said to threaten some of the central functions of the body, this could perhaps be said to be the case if we are considering castration (in particular physical castration).

characterizes TMS is that it causes an effect on the brain by the use of a magnetic field that passes unimpeded through both the skin and skull. However, it is clear that all four methods—including TMS—can be said to infringe on bodily integrity in the interpretation of causing a non-consensual change *in* the body. Therefore, on the final two accounts it seems that the plausibility of premise 1) can be upheld: compulsory neurointerventions can be said to intrude on a person's bodily integrity. However, this explication of the ways in which this will happen serves as a nice point of departure for considering the plausibility of premise 2); this is whether non-consensual effects on bodily integrity violate a right and, therefore, must be regarded as morally wrong.

Is it plausible to hold that individuals have a right to the protection of their bodily integrity in either of the final two interpretations which may be relevant in relation to neurointerventions? It will be appropriate to address this question with a simple methodological comment. A certain moral right is often theoretically defended by showing that the existence of this right makes it possible to pinpoint and explain a number of acts that strike us as morally wrong. We can explain that acts a, b, and c—which all seem wrong—are in fact unacceptable because they violate this particular right. However, this way of sustaining the existence of a right is not in itself sufficient as a justification. If it turns out that some other acts d, e, and f, which all seem morally unproblematic (or even desirable), will be comprised by the right, in the sense that a person who carries out these acts will end up as a right-violator and, hence, will be acting wrongly, then this seriously challenges the claim that there exists such a right (see e.g. Ryberg 2008).[26] This is a rather trivial point. However, it is highly relevant when it comes to the assessment of the right to bodily integrity as explicated in interpretation (F). Insisting that there is a right to bodily integrity in the sense that someone violates this right by causing non-consensual changes in another person's body seems highly questionable. It is correct, of course, that a forced neurointervention

[26] To avoid any misunderstanding, I am not holding that a moral theory cannot plausibly imply that some acts which we usually tend to regard as unproblematic are in fact wrong. However, without going into too much detail, I do believe that the more a theory happens to be in conflict with what strikes one as morally right or wrong, the better are the reasons it will have to advance in support of its case. However, it is remarkable how little adherents of the right to bodily integrity have had to offer in this respect and, more importantly, it is difficult to see what a justification would amount to. For instance, arguments based on the more basic views that 'you are your body', that your body is necessary in order to pursue important goals, or that it is required in order to exercise autonomy do not seem to underpin a right to bodily integrity in this interpretation (nor, for that matter, in any of the others). Intrusion of bodily integrity need not conflict with any of these more basic values. For an illuminating critical discussion of possible justifications in relation to the libertarian idea of self-ownership, see Lippert-Rasmussen (2008).

will result in a certain physical change in an offender's brain. However, it is also clear that we often cause non-consensual effects in other people's brains by ways of acting that cannot plausibly be held to involve violations of a right. This is the case in various interactions with other people. If I ask my daughter to clean her room, the hearing of this question involves an effect in her brain and my interference may well be unwelcome, but few would hold the posing of this question to be morally unacceptable. Similar interferences happen all the time in ordinary life as well as in criminal justice contexts. A defendant may be asked various questions which he would rather do without. He may have to listen to testimonies that he would prefer not to hear. He may have to see things he would prefer not to see. And, if he is found guilty, he will have to go through various unwanted experiences. All such interactions involve non-consensual alterations of neural processes in the brain. But they are not morally unacceptable.[27] Thus, a right to bodily integrity, understood as a right that protects an individual against interferences resulting in non-consensually caused brain processes, will turn numerous reasonable interactions with others into violations of this right.

It might perhaps be held that this way of rejecting a right protecting against non-consensual changes in the body is premature. If one subscribes to this interpretation of the right, what one is concerned about is not mere changes at a molecular level—as caused in the outlined examples—but more substantial morphological alterations in the body. However, this specification of interpretation (F) has little to offer. First, it is a fact that long-term influences on the brain may cause more substantial morphological or structural effects. But this means that, for instance, in the same way as studies have indicated that years of navigation by London taxi drivers correlate with an increase in hippocampal grey-matter volume followed by a corresponding decrease after retirement, it may well be the case that the brain of inmates, who are being deprived of their usual pre-punishment practices and stimulations—e.g. like a taxi driver who can no longer work because he is placed behind bars—may undergo more substantial volume alterations in the brain (see e.g. Maguire et al. 2006; Woollett and Maguire 2011). Second, and more importantly, this qualification of the interpretation suffers from the obvious problem that

[27] To suggest that all these ways of affecting other people's brains are fully acceptable, but that compulsory neurointerventions are not, because only the latter sort of interference involves an intention of causing changes in a person's brain, is of little help. In neither of the two cases is the alteration of the brain intended as an end in itself (the purpose of the neurointervention is a change of conduct) and, in so far as the alteration of the brain is held to be intended as a means to the desired end, this means can be held to be equally intended in both cases.

premise 1) will, for several types of neurointervention, no longer be true. For instance, pharmacological treatment or TMS seem to cause alterations in the brain only at a microscopic level.[28] Thus, the qualified right will no longer protect against such interventions. Thus, all in all, the objection against non-consensual neurointerventions on offenders based on interpretation (F) does not seem convincing.

But what if we consider instead the narrower account of bodily integrity (E) concerning penetrations of the body? Is it not at least plausible to contend that people are in this sense morally protected? In contrast to interpretation (F), it does not seem to be the case that we are often subject to such non-consensual bodily intrusions in ways that are morally unproblematic. Thus, this account of the right may seem more promising as an objection against compulsory neurointerventions. However, in considering this right, an obvious question concerns the stringency of the protective force that is provided.[29] To adopt an absolutist position, that is, to hold that infringements of this right can never be morally acceptable, is not convincing. That even a minimal intrusion into the body should be unacceptable, even if this constitutes the only way to prevent significant countervailing reasons, is hard to believe. A more promising view, I suggest, is to defend a threshold position, according to which the right can be overruled when enough is at stake. However, the most plausible version of a threshold approach seems to be one that determines the threshold relatively to the extent to which the right in question is being infringed.[30] If such a view is correct, then this is interesting in relation to our discussion of compulsory neurointerventions.

Given the interpretation of bodily integrity we are here considering, it seems reasonable to hold that those neurointerventions that do infringe on a right to bodily integrity will do so only to a very modest extent. This is most obvious with regard to the possibility of being forced to take a pill. Surely, this must—whatever the proper parameter for the measurement of intrusions—be regarded only as a minimal intrusion of integrity. A pill enters the body by being swallowed (or through the gastrointestinal wall)

[28] To suggest that an intrusion on bodily integrity does not include simple transient changes in the brain, but only permanent changes, is of little help in the present context. As previously noted, most of the relevant neurointerventions do not cause permanent changes.

[29] Another type of consideration on the moral force of the right to bodily integrity has been presented by Douglas (2014).

[30] For instance, see Kagan's considerations of thresholds in relation to the constraints against harm. As he remarks: 'it seems as though a great deal must be at stake before it would be permissible to cause someone excruciating physical pain; but inflicting some mild discomfort would presumably involve a far lower threshold' (Kagan 1998, p. 82).

in precisely the same way as the food we eat. While many rights protect us against intrusions that from the outset we regard as highly undesirable, this does not seem to be the case with regard to this particular intrusion which, after all, is an activity which we perform unproblematically in our everyday life. Obviously, this is not tantamount to holding that it is unproblematic to be forced to such an activity. But the undesirability of the forced activity seems to consist almost exclusively in the fact that one is being forced, that is, that one is being deprived of self-determination or the possibility of exercising autonomy, rather than in the inherent undesirability of the activity itself (i.e. the bodily intrusion). However, what we are considering here is not the exercise of autonomy, but the nature of the bodily invasion and, as such, being forced to take a pill must be regarded as a minimal intrusion.[31] The same, I suggest, is the case when it comes to neurointerventions in the form of injections. Even such procedures seem to constitute only very modest ways of invading the body. It is difficult to imagine a less comprehensive way of penetrating the boundary of the body. It can be conducted almost insensibly, it is highly transient, and leaves the body boundary unmarked. Moreover, it is worth noting, almost identical compulsory intrusions have been conducted in a number of cases without much resistance. For instance, even though a warrant is required in some jurisdictions, it has for many years been standard practice to require blood samples of motorists arrested for drink driving. And several countries have had programmes involving the compulsory taking of blood samples from prisoners in order to test for infectious diseases. Few have objected to such practices merely on the grounds of the magnitude of bodily intrusions. But, if this is correct, then it seems plausible to suggest that the thresholds for such infringements of the right to bodily integrity must be set at a rather low level and, consequently, that it is far from clear that such infringements are unacceptable when they are balanced against what will be gained if a treatment succeeds in preventing criminal behaviour.

The picture, of course, is very different with regard to surgical interventions in the brain. Such a treatment clearly constitutes a much more comprehensive intrusion into the body and, consequently, may well be held to require much stronger countervailing reasons in order to be justified.

[31] Obviously, the effects of the treatment may not be insignificant. On the contrary, they may result in a significant change in the mental life of the offender. However, this is a type of effect that reaches beyond mere bodily intrusion. The moral significance of such effects will be considered in the ensuing section.

However, it does not strike me as highly controversial to contend that even more comprehensive intrusions can sometimes be justified. For instance, analogously, suppose that the only way the police can prevent an offender from killing or seriously harming another person is by shooting the offender in the leg (or another non-vital part of the body). Such interactions by the police are accepted in all jurisdictions as a last resort procedure, and it does not seem absurd to regard such a practice as morally acceptable. But if that is the case, then why should it not in some cases be acceptable to conduct a surgical procedure on an offender if this would constitute the only way to prevent a future serious crime? Obviously, it will rarely (if ever) be possible to predict that a crime can be prevented by a treatment with the same degree of certainty as when the police faces an offender *in flagrante delicto*. But if hypothetically this were possible, then it is not clear that there is a significant difference *in terms of bodily intrusion* between being shot at and being forced to undergo a surgical procedure. In other words, even more comprehensive interventions into the body may sometimes be justified.[32]

Where does this thus lead us? I believe the upshot of the previous discussion lies in the following. What we have considered is an objection, namely that compulsory neurointerventions should be rejected because they violate an offender's right to bodily integrity. However, on closer scrutiny, this objection turned out to be less convincing. First, we have seen that the concept of bodily integrity requires clarification and that in several of the suggested interpretations there were no reasons to believe that a neurointervention would in fact intrude on an individual's bodily integrity. That is, premise 1) of the objection would be false. Second, there were still two interpretations that seemed pertinent in relation to neurointerventions. However, the first interpretation (F), associating intrusions on integrity to non-consensual bodily alterations, did not provide a plausible background for a moral right the violation of which should be regarded as morally unacceptable. That is, in this interpretation, premise 2) of the objection became false. The second interpretation (E), relating intrusions on integrity to incursions into the body, seems to constitute a more promising candidate for a moral right. However, even though surgical procedures must clearly be regarded as comprehensive incursions on the body, the other categories

[32] If one believes that it would be justified for the police, as a last resort, to shoot a person—who is about to kill someone—in the head, then the parallel to neurosurgery obviously becomes even closer; that is, it will become difficult to explain why such neurointerventions unacceptably violate bodily integrity.

of neurointervention—presumably constituting the most realistic types of treatment—seem only to involve very modest intrusions on bodily integrity in this particular sense. In other words, it does not appear implausible to believe that a right protecting bodily integrity in such cases could be overridden by what may be gained in terms of crime prevention. In fact, it was indicated that this may be the case, even if we consider surgical interventions into the body. Thus, even though it would be premature to conclude that sufficient reason has been provided for rejecting the relevance of a right to bodily integrity, I still believe that the previous considerations—underlined by the fact that TMS did not at all constitute an intrusion of bodily integrity in the most pertinent interpretation—indicate that if one worries about the use of neurointerventions as part of mandatory treatment schemes, the main concern has little to do with the direct impact on the body, but rather with some of the *effects* caused by such physical interventions. What this means, is the question to which we shall turn in the following section.

The Right to Mental Non-interference

Consider again the previously mentioned examples of aversion therapy conducted in the 1950s and 1960s. Why do these types of treatment strike us as highly unacceptable? There are probably several reasons. As we have seen, one obvious answer is that treatments were imposed on individuals who ought not to be treated. In cases where the individuals were referred by the courts, they were convicted of behaviour that ought not on any plausible theory of criminalization to be legally prohibited, namely homosexuality and transvestitism. However, this is only part of the explanation. Even if such treatment methods are imposed as reactions to conduct that definitely ought to be criminalized—such as the brutal violence carried out by Alex in our fictional example—they would still strike many as unacceptable. Moreover, this probably has little to do with the stringency of the direct physical effects caused by the treatment (even though, as mentioned, some did involve electric shocks). What mainly drives the dissociation to these types of treatment, I imagine, is the fact that they were designed to alter the mental life of those treated.

That a corresponding judgment may lie behind much of the anxiety against the imposition of neurointerventions as a way of avoiding recidivism seems a reasonable belief. In fact, not only does the idea of forcibly altering an offender's mind give rise to immediate concern, this particular aspect of invention has by several theorists been underlined as constituting, on reflection, the main ethical problem associated with such mandatory treatment schemes. For instance, Jan Christoph Bublitz holds that 'we would close our eyes to the real interference to which neurorehabilitation amounts if we remained on the evident, external and bodily level and were concerned about pinpricks only. Neurorehabilitation primarily alters minds' (Bublitz 2018, p. 299). Likewise, Jared N. Craig contends that 'there are strong reasons to believe that the most compelling arguments against non-consensual DBIs [direct brain interventions] do not rest on the right to bodily integrity, but on the more fundamental right to "mental integrity"' (Craig 2016, p. 111).

There are conceptual differences in the way this concern has been formulated. Some theorists, as just indicated, talk of a right to mental integrity, while others have defended the idea of a right to mental self-determination. Yet others have stressed the significance of a right to freedom of thought.[33] Since the different ways of presenting the idea behind the anxiety are not necessarily synonymous—for instance, presumably, freedom of thought constitutes a narrower right than the right to mental integrity—I will here talk broadly of a *right to mental non-interference*. Given this way of phrasing the idea, the objection that we will now consider is that compulsory neurointerventions are morally unacceptable because they violate the offender's right to mental non-inference. Parallel to our discussion of bodily integrity, this conclusion will in the simplest form be based on the premises that: 1) compulsory neurointerventions infringe an offender's right to mental non-interference and that 2) an infringement of the right to mental non-interference is morally unacceptable. Obviously, the cardinal issue in the assessment of this objection is how the concept of mental non-interference should be understood. As we will see, it is not a simple task to give content to this concept in a way that renders the objection valid. From the outset, I believe that three overall interpretations of the right to mental non-interference can be presented; these will be considered in turn. As will be argued, the different interpretations face serious problems when it comes to the satisfaction of the conjunction of the two premises.

[33] For instance, Boire has held that 'compelling a person to use a pharmacotherapy drug not only implicates the person's traditional rights to bodily integrity and informed consent, it also implicates the fundamental rights to freedom of thought' (Boire 2010, p. 288).

(A) The first and simplest view is to hold that a right to mental non-interference comprises all types of mental phenomena. In this very broad interpretation, the right morally protects an individual against any interference which non-consensually influences conative, cognitive, and affective elements of our inner life. Given the complexity of our mental life, and the various types of semi- or subconscious components of the mind, the protective zone provided by the right becomes very wide. This has the obvious advantage that premise 1) becomes true. No matter whether neurointerventions affect the contents or capacities of the mind, or whether they cause effects pertaining to conative, cognitive, or affective aspects of our mental life, such interventions will, if non-consensually imposed, infringe the right to mental non-interference. However, for the opponent of neurointerventions this advantage is dearly bought.

The obvious problem is that on this account of what the right protectively comprises, premise 2) becomes highly implausible. In the same way as a right protecting against alterations inside the body was implausible because such alterations—if only at a microscopic level—are caused all the time in the interaction with others, in a totally unproblematic way; the right protecting against all sorts of non-consensual mental influence will become equally over-inclusive. That an unwanted question posed by a judge, initiating associations, provoking memories, or causing emotional reactions in a defendant, should imply that something morally wrong has happened is clearly absurd. In a similar manner, but more generally, such a right would turn numerous types of interaction with others in our daily life into morally unacceptable behaviour. And, to mention a final implication, this right would probably, due to the unwanted mental side-effects, be incompatible with all types of punishment (as noted above, incarceration has various types of psychological effects and even a fine may cause an unwanted emotional reaction). It is of little help to add the further qualification that a right violation not only requires a non-consensual influence on the mind of someone, but also that the mental effect caused is in fact intended. Surely, the posing of the judge's question cannot constitute a right violation even if it is borne by the intention of prompting a memory in the defendant. Thus, this broad account cannot constitute a plausible interpretation of a right to mental non-interference.

(B) Another, and *prima facie* more plausible approach, would be to specify more narrowly the parts of our mental life that the right to mental non-interference is supposed to protect. The obvious challenge, of course, is to indicate what this delimitation should consist in. What parts or aspects of our mental life require a particular moral protection? The most obvious suggestion, I imagine, is to hold that an individual should be protected against behaviour that will have deleterious effects on the mind. There are various types of mental states that we would, at least under usual circumstances, very much prefer to be without. For example, the feeling of pain. Inflicting pain on others constitutes the standard example of a wrong. Another example could be the state of fear. It certainly does not seem unreasonable to hold that I have been wronged if someone does something to seriously frighten me. Yet other examples could be to place another in a state of sorrow, anxiety, frustration, sadness, etc. Obviously, the task of specifying precisely what should be regarded as deleterious mental effects may not be an entirely simple task. And, one may be confronted with the problem that there could still be some situations in which it does not seem wrong to cause such emotional reactions, even where they are unwanted. However, setting these issues aside, it seems generally reasonable to believe that by delimitating the parts of our mental life along these lines, it becomes more plausible to appeal to a right to mental non-interference and to hold that premise 2) must be true. But on closer inspection it is clear that this cannot be the case.

The obvious problem again is that even this explication seems over-inclusive, for the simple reason that standard types of punishment usually cause all the specified types of emotional reactions in offenders. As mentioned in the above discussion of the identity objection, there is a growing research establishing the various mental implications of incarceration (see Schnittker 2014; Turney et al. 2012). In fact, recent research has even documented that other types of criminal justice contact—such as arrest and conviction—are deleteriously associated with mental health (see Sugie and Turney 2017). Thus, given this specification, it seems that one cannot at the same time consistently subscribe to the use of imprisonment as an acceptable type of punishment.

But how should we then specify the parts of our mental life which the purported right is supposed to protect? Another way of cutting the cake would be to focus on particular mental *capacities*. Such an approach has been presented by Bublitz. His main concern is not to consider what type of influence on the mind should be regarded as morally right or wrong but to outline the contours of a theory of which aspects of the mind should be *legally* protected. However, his considerations could be held to apply with equal force to a purely normative evaluation. What he defends is a right to 'mental self-determination' comprising (partly) the capacities by which we can be said to control our mental life. As he puts it: 'persons can be described as having some sort of conscious control over their mental life. Interventions can undermine these control capacities and thereby interfere with the right to mental SD [self-determination]' (Bublitz forthcoming, p. 107). It is hard to dispute that we seem to possess several such capacities. For instance, we can decide to engage in contemplation of a particular issue, to focus our mind in a particular way, to try to recollect something that is deeply hidden in our memory, to shift attention, to pull ourselves together, etc. All such capacities can reasonably be regarded as ways of exercising mental control. And, it has immediate appeal to suggest that non-consensual deprivation of any of these highly important capacities would be morally dubious. Thus, does a control-oriented interpretation of the right provide a more solid base for an objection against forcible neurointerventions?

It seems to me that the main problem confronting this explication of the objection is that it is far from clear that non-consensually imposed neurointerventions work by undermining an offender's control-capacities. First, mental phenomena that are being controlled may well be altered without causing an effect on a capacity. As a simple analogy, if I possess the capacity for mental arithmetic, this capacity is not in itself affected by the fact that it is applied to calculating some figures and not to others (and this is so, even if the other figures hypothetically enter my mind as a result of some sort of unwanted external stimulation). Correspondingly, insofar as a neurointervention functions by affecting the content of our mental life, say, that it alters a set of preferences that drive criminal activity, there is no reason to believe that an offender's *capacities* for controlling his mental life will be affected. Second, if a neurointervention affects an offender either by reducing the force of a strong preference (perhaps, a sexual desire) by increasing certain cognitive functions (say, the ability to comprehended more distant implications of an act) or by increasing an offender's ability to restrict the influence of certain immediate impulses (perhaps, for aggressive behaviour) then either the individual's control capacities will remain unaffected or the possibility of exercising proper control will be increased. Thus, in this interpretation, it is far from clear that premise 1) is true (see also Petersen and Kragh 2017).

Another attempt at providing a proper delimitation could be to contend that a right against mental non-interference concerns aspects of our mind which are particularly important; for instance, which serve a crucial role in who we are or in the life we lead. If an aspect of our inner life constitutes a core element of our mental set-up, then it seems more appealing to hold that it calls for a particular moral protection. However, very little work has been done to specify what should in the relevant sense be regarded as particularly important features of our mental life.[34] Thus, it still remains to be shown that it is possible to specify these aspects of our inner life in a way that implies that (some) compulsory neurointerventions infringe on the right to mental non-interference, and that this infringement can be deemed morally unacceptable without, at the same time, turning something that we accept as unproblematic interaction between people—or morally required punitive reactions to criminal offending—into wrongful behaviour. For instance, as we have already seen in the previous discussion of the identity objection, the view that it is unacceptable to influence someone's personality may not be consistent with the acceptance of long-term incarceration of offenders.

(C) A third approach might be to change the focus from more narrowly specified influences on the mind to the way in which effects on the mind are caused. Following this idea, what the right to mental non-interference protects an individual against is not particular alterations of the mental set-up, but rather the influencing of a mind in particular ways. For instance, it might be held that, while indirect influences are irrelevant from a rights perspective, the right to mental non-interference protects against influences on the mind by the use of methods that *directly* affect the chemistry or physics of the brain. Another possibility would be to distinguish between influences that pass through the perceptual system and those that do not, by contending that the latter but not the former types of influence are protected by

[34] To my knowledge, Bublitz is the theorist who has done most to defend the significance of a right to mental self-determination. However, though he is explicitly aware of the delicate balance between specifications that are over-inclusive and under-inclusive, his considerations of what the right is supposed to protect leave, in my view, a somewhat imprecise picture. For instance, in one place he holds that 'mental SD [self-determination] protects against interventions that undermine control capacities or change elements of the personality' (forthcoming, p. 108) and, furthermore, explains that he believes that the personality of an individual can be broken down into 'a web of desires, beliefs, emotional propensities, etc.' (p. 106). In another place he (along with Merkel) suggests that 'worrisome interventions are those that reduce or impair cognitive capacities (e.g. memory, concentration, willpower), alter preferences, beliefs and behavioural dispositions (e.g. implanting false or erasing true memories, creating addiction), elicit inappropriate emotions (e.g. artificially induced appetite) or clinical identifiable mental injuries' (Bublitz and Merkel 2014, p. 68). He also claims that the 'rights to mental self-determination (SD) should encompass all mental states, i.e. not only thoughts or opinion, but also emotions, behavioural dispositions, even subconscious processes' (Bublitz forthcoming, p. 106). Given the latter quote, it will be very hard to avoid the problem of over-inclusiveness leaving premise 2) implausible. And the earlier quotes leave open questions, such as whether only the desires and emotions that can be regarded as part of the personality (and which would undermine it if they were changed), are comprised by the right, or whether all desires and emotions are. Moreover, it remains somewhat unclear whether it is only deleterious alterations of capacities or emotions, or any type of alteration, that conflicts with the right.

the right. Whether these two suggestions amount to different interpretations of the right to mental non-interference or whether they coincide obviously depends on how the distinctions are more precisely expounded. However, I tend to believe that the distinctions are usually used synonymously or co-extensionally.[35] Be that as it may, what matters is that it is the way an individual is affected, not the result of an intrusion, that may infringe on the right and that this way of understanding the right has some immediate appeal. For instance, as noted at the beginning of the chapter, Beehler specifically underlines that if one imagines that Alex's mind were to be reformed not by applying the Ludovico Technique—which operates through the senses—but through some sort of brain surgery, then we would be inclined to regard this as an even more radical intervention, even if the resulting effects on his mental life were precisely the same. Furthermore, notably, if it is plausible to hold that there are certain *methods* of influencing the mind that are morally unacceptable, then not only will premise 2) be true, but the same may well be the case for premise 1). The problem confronting some of the earlier interpretations, namely, that it did not seem to be the case that neurointerventions would actually infringe on the right to mental non-interference, would apparently be avoided. What characterizes neurointerventions precisely is that they circumvent the perceptual system, that is, that they operate directly on the brain. Thus, does this method-based account constitute a more plausible interpretation of the right to non-interference and, hence, of the objection against compulsory neurointerventions?

In my view the answer must be in the negative. On closer inspection, it seems to me that what gives moral credence to this interpretation is not in itself the distinction between indirect and directs ways of affecting the mind or whether or not an effect is caused through the sensorial apparatus, but rather something that is only contingently associated with these distinctions. If one asks whether it makes a difference if someone tries to influence our minds either by the use of a method that leaves room for resistance or in a way that does not, then the question will usually be answered in the affirmative. For instance, this seems to be what drives part of the intuition when Farah in an oft-quoted passage contends: 'In anger management class, a person is free

[35] For instance, Bublitz and Merkel hold that 'indirect (or external) interventions are those stimuli which are perceived sensually (i.e. heard, seen, smelled, felt even if not apprehended or reflected upon consciously) and pass through the mind of the person . . . By contrast, direct (or internal) interventions are stimuli reaching the brain by other routes than sensual perception' (Bublitz and Merkel 2014, p. 69).

to think, "this is stupid. No way I am going to use these methods." In contrast, the mechanism by which Prozac curbs impulsive violence cannot be accepted or resisted in the same way' (Farah 2007, p. 28). Along the same lines, Bublitz holds, as part of his attempt at developing a legal framework for the protection of the mind, that the guiding idea of the proposal is that persons have more control over some interventions than over others (Bublitz forthcoming, Chap. 4, p. 116). And that 'persons have most control over indirect interventions whose sensual substrates they perceive' while direct interventions, by contrast, seem to circumvent a person's control mechanisms (Bublitz and Merkel 2014, p. 68f; Bublitz forthcoming, p. 112ff). However, this way of buttressing the moral significance of the direct/indirect or the perceptual/non-perceptual distinctions is not persuasive.

First, though this has often been ignored in the discussion of neurointerventions, it is obvious that there are some interventions which are forcibly imposed, which operate directly on the brain, but which are nevertheless fully controllable by the individual who is the subject of the intervention (see Ryberg 2015a). Suppose, for instance, that DBS micro-electrodes were implanted in the brain of an offender as part of a compulsory treatment programme designed to counter explosive aggression. Suppose, further, that it is possible for the offender—as is the case in many Parkinsonian patients with implants—to switch off the signal from the electrodes or control the amplitude if he wishes. In this case, the intervention will operate directly on the brain, but the offender will have full control over the possible effects on his mental life.

Second, it does not seem plausible to contend that the effects of indirect interventions that are mediated through our sensual system are always controllable. On the contrary, there is today a comprehensive amount of research establishing how both our judgements and behaviour are unconsciously affected by stimuli that pass through our sensory systems. For instance, a picture of a pair of eyes has been shown to unconsciously affect the likelihood that people will remember to put a few coins in an honesty box as payment for the milk they use in their coffee (Bateson et al. 2006; Nettle et al. 2013). Other unconscious effects have been shown to follow from stimuli operating on the olfactory system. For instance, in one study subjects were asked to consider morally a number of different actions and assess them on a Likert scale. It turned out that when subjects made the assessments under the influence of a highly unpleasant smell they became significantly tougher in their ethical judgments (Schnall et al. 2008).

Corresponding studies have been conducted involving influences on the gustatory system. When people have a bitter taste in the mouth while engaging in ethical evaluations, they tend to regard different types of behaviour as more wrong (Eskine et al. 2011). And these sensual effects on judgment occurs unconsciously. Similarly, as Daniel Kahneman has noted as part of the explanation of numerous types of priming effects revealed in recent psychological studies: 'An idea that has been activated does not merely evoke one other idea. It activates many ideas, which in turn activate others. Furthermore, only a few of the activated ideas will register in consciousness; most of the work of associative thinking is silent, hidden from our conscious selves. The notion that we have limited access to the workings of our minds is difficult to accept because, naturally, it is alien to our experience, but it is true' (Kahneman 2011, p. 52). In other words, the view that effects on the mind that follow from stimuli through our sensual system are controllable is of course sometimes true, but quite often it is not. But if the possibility of exercising control or resistance does not follow the dividing line between direct and indirect or perceptual and non-perceptual ways of influence on an individual's mental life then, in the absence of any other morally significant factors—and indeed it is hard to think of any that non-contingently follow the distinctions—it becomes arbitrary to maintain the view that one type of influence should be regarded as morally unacceptable while the other is not.[36]

Are there any other ways in which content can be given to the view that the right to mental non-interference does not protect against particular effects on the mind but rather against ways of causing the effects? The only other method-based answer I can think of has been tentatively put forward by Craig as part of his defence of a right to mental integrity. He suggests that the 'speed' in which a neurointervention affects human agency may be an

[36] It might be suggested that there is another way in which one could be said to possess control when stimuli affect us through the sensory system, rather than when our brain is affected directly: One could decide to prevent the access through the senses. For instance, even if an unconscious visual effect on our mind is in one way uncontrollable, it would still be possible to resist it if one decides to close the eyes. However, this is not a reasonable way of accounting for the perceptual/non-perceptual distinction. First, in practice it will be possible for us to close our senses to the external worlds (e.g. you cannot avoid olfactory or gustatory experiences). Second, if one insists that the cutting off of the senses is at least *in principle* a possibility, then the same can be held with regard to non-perceptual (or direct) interventions. For instance, an individual who is the subject of a non-consensual direct brain intervention as part of a treatment for aggression, can—in principle—try to prevent the effects by taking drugs that increase the level of aggression or by engaging in other attempts at boosting his aggression (at the end of the day, suicide is also a possibility). Thus, neither in practice nor in principle does this proposal seem plausible.

ethically relevant feature (Craig 206, p. 115). However, this does not seem correct. Obviously, there can be cases in which a fast alteration of the mental set-up of offender is less desirable than a slower alteration. But there is no reason to believe that this is necessarily or generally the case.[37]

In summary, what the previous considerations show is that it is not a simple task to give content to the idea of a right to mental non-interference that makes the objection against compulsory neurointerventions plausible. Several of the considered interpretations of the right were confronted with the challenge of over-inclusiveness, which implied that premise 2) of the objection became implausible. In other interpretations, it became more reasonable to regard premise 2) as true. However, with these explications it was not the case that compulsory neurointerventions would involve an infringement of the right; i.e. premise 1) would be false. Thus, an explication of the right that succeeds in providing a more general rejection of the compulsory use of neurointerventions by satisfying the conjunction of the two premises has, in my view, not been provided.

Finally, it should be mentioned that there is another challenge that relates to the acceptance of premise 2), concerning the stringency of the moral protection provided by the right to mental non-interference. Parallel to the discussion of the right to bodily integrity, it is hardly plausible to present the right to mental non-interference as an absolute right. In fact, several adherents of the view have explicitly held that such a right should not be interpreted in absolutist terms. For instance, Craig has outlined the overall contours of the view that the right to mental integrity should be balanced against 'interests in crime-prevention and rehabilitation' (Craig 2016, p. 13). Craig is explicit about the fact that he has presented only a rough outline of the idea of a case-by-case assessment, and it remains unclear what he believes constitutes the criterion of rightness when the balancing is conducted. But the overall point is that a fully fledged account of the objection considered in this section, will have to include some considerations establishing that the interests of potential crime victims are insufficient to warrant infringements of the right to mental non-interference.

[37] As Petersen and Kragh have put it: 'it is far from obvious why we would wish a violent person to have a slow recovery (instead of a speedy recovery) and rehabilitation especially if the intervention . . . has been proven to be safe and effective' (Petersen and Kragh 2017, p. 32).

Conclusion

This section has dealt with some of the ethical concerns pertaining to the use of neurointerventions as part of compulsory treatment imposed upon offenders. More precisely, we have considered whether there are more general or in-principle reasons for ruling out such interventions as acceptable methods of reducing the risk of recidivism. First, we have considered part of the more general debate on moral enhancement. The point of doing so has not been to suggest that compulsory neurointerventions are rightly placed in this conceptual category. Nothing hinges on whether such interventions are perceived as *moral* enhancements, or whether they should be regarded as *enhancements* at all rather than as methods of treating individuals in order to reduce criminal proclivities. The point simply was that some of the concerns advanced in this debate *prime facie* seemed to apply equally to the forcible use of neurointerventions. However, as argued 1) neither reasons based on the lack of a proper moral yardstick, on the freedom to fall, nor on the effects of the identity of those who are treated, seemed to provide a general background for a rejection of compulsory neurointerventions. Secondly, we considered the significance of an objection based on the right to bodily integrity. It was argued that 2), forcible neurointerventions either did not infringe on this right at all or they did infringe on this right, but only to a very modest extent. Furthermore, it was indicated that even more comprehensive infringements of this right could, under some circumstances, be justified. Finally, I have considered the most obvious candidate for a right protecting against compulsory neurointerventions, namely, a right to mental non-interference. It was argued that 3), an objection based on this right faces problems when it comes to the task of establishing that forcible neurointerventions are unacceptable without ending up as being unacceptably over-inclusive in a way that undermines the plausibility of the purported right.

Given these conclusions, it may be appropriate to repeat the initially underlined point that there are, of course, various ways in which everyone would agree to regard such neurointerventions as wrong. There may well be some cases in which interventions have no or only a marginal effect, or even when there is a desired effect this may be accomplishable in ways that are less intrusive or in other ways preferable. Furthermore, even if such alternatives do not exist, the consequences of imposing such compulsory measures may not be preferable to what is gained in terms of reduced risk of recidivism. However, I take it that many readers might well feel that such empirically contingent considerations do not really exhaust the ethical discussion of this admittedly highly controversial issue. That is, that even if the conclusions of this chapter are valid, there may well be other more general or in-principle reasons for rejecting compulsory use of neurointerventions on offenders. Some further candidates for such objections naturally arise and will be considered in relation to the (not less controversial) topic to which we will turn in the next chapter.

4
Neurointerventions as Punishment

The theoretical discussion of the use of neurointerventions on criminal offenders, as the considerations in the previous chapters indicate, has mainly dealt with the question as to whether coercive or compulsory use of such interventions can be ethically justified. The focus has been placed on criminal offenders for the obvious reason that prevention of future crimes constitutes a highly desirable goal and that there is sometimes a risk that offenders may fall back into criminality. However, the fact that offenders are serving a punishment has often had very little significance in the discussion. Beyond the purely theoretical parallel between punishment and forced neurointerventions, and the more practical considerations of the fact that penal reductions can be used as incentives for accepting participation in treatment programmes, the closer relation between a treatment by neurointervention and the punishment imposed on the offender has often remained somewhat unclear.

In other words, there has been a clear tendency to address the question of the ethical legitimacy of the use of neurointerventions as either a problem within medical ethics or as a special instance of a problem in the enhancement debate (see Ryberg 2015b). Comparatively, penal theoretical considerations have played a much more modest role, if any role at all. However, in this chapter the perspective is changed. More precisely, the purpose will be to engage in penal theoretical considerations in order to answer the question as to whether a forced neurointervention could in itself function as a type of *punitive* reaction to criminal conduct. But why, one might ask, is this question interesting? If one is interested in whether compulsory use of a neurointervention can ever be justified as a way of preventing re-offending, why then push the discussion a step further by asking whether such an intervention can also serve as a punishment? A simple answer is that a retributively-minded commentator might respond to the discussion by underlining that the purpose of the criminal justice system is to punish offenders, not to engage in treatment.

Neurointerventions, Crime, and Punishment. Jesper Ryberg, Oxford University Press (2020). © Oxford University Press.
DOI: 10.1093/oso/9780190846428.001.0001

This objection makes it important to consider whether a neurointervention could possibly function as a punishment. However, there are further reasons for addressing this question. Here are some.

First, as we have seen, the discussion whether it is morally acceptable to offer penal reductions as a way of persuading offenders to take part in treatment programmes has usually focussed on whether such offers are unacceptably coercive or otherwise inappropriate. However, as earlier argued, it is clear that there is also a penal theoretical side to this discussion. For instance, such reductions would in themselves be unacceptable if seen from the perspective of a positive retributivist view, according to which an offender should receive no more and no less than the punishment he deserves. Correspondingly, there is a penal theoretical side to the idea of using penal reductions as a sort of compensation to offenders who are forced to participate in treatment programmes. However, if a neurointervention could be imposed as a type of punishment, then the picture changes significantly. For instance, the fact that an offender gets two years in prison rather than three, if he accepts or is forced to go through a treatment, need not constitute a penal theoretical problem. If, hypothetically, the imposed treatment per se constitutes a punishment corresponding in its degree of severity to about one year in prison, then the treated offender may all in all end up receiving the appropriate punitive response to the crime committed. From having constituted a problem in terms of disproportionate punishment, we are instead left with a question concerning the imposition of one type of punishment rather than another which, from a retributivist perspective, may be much less problematic. (In fact, as we shall see shortly, the existence of such penal alternatives may be highly desirable for the retributivist.) In more general terms, if there are certain advantages related to a scheme where the neurointervention is imposed in combination with a change in the standard punishment which an offender receives, then it may make a significant difference for the overall ethical assessment of this scheme, as to whether or not the neurointervention can in itself function as a type of punishment.

Second, as we have seen, a major objection to offering penal reductions as incentives to participation in treatment programmes and, not least, to the use of compulsory treatment schemes, has been that such schemes involve a violation of the requirement of consent. There could be reasons for holding that, even though informed consent constitutes an imperative in standard medical treatment, this need not be the case if one considers such a special issue as the use of neurointerventions on criminal offenders as a means to preventing future crimes. But, even in this context, many would probably be highly reluctant to abandon the consent requirement. However, as argued, if a neurointervention could be imposed not as a treatment but as a type of punishment then it becomes much harder to maintain this position. No one would insist that a punishment requires informed consent by the offender. Thus, the discussion of the significance of consent in itself makes it relevant to consider whether neurointerventions can serve genuine punitive functions.

Third, as parts of the previous discussion, as well as the general debate on neurointerventions, have shown, there are some objections to the compulsory use of neurointerventions on offenders that cannot be upheld if one at the same time accepts that offenders should be punished for their misdeeds. However, as we have also seen, the fact that a punishment constitutes a compulsory reaction does not in itself imply that compulsory neurointerventions are acceptable. But, from the outset, it is natural to reflect on the parallels between these two types of compulsory procedure. It seems obvious to ask how it can possibly be justified to place an offender in prison for many years but not to perform a physical intrusion such as giving a few forced injections. Or, how acceptance of the long prison term is consistent with a rejection of a compulsory, but minor and transient, alteration of an offender's impulse control. Part of the answer to such questions could be that a neurointervention involves a wrong that is not involved in the prison term. Answers along this line were considered in the previous chapter. However, as noted, another answer would be that punishment is something special. But then it becomes obvious to ask what it is that makes a punishment morally different from other types of compulsory treatment. And one way of considering this question is to ask whether it is possible that neurointerventions could function as punishments.

Fourth, as repeatedly underlined, the whole point of considering the use of neurointerventions on criminal offenders is that such interventions could constitute a way of preventing future crimes. But this means that decisions on whether a certain treatment should be effectuated depends upon the possibility of estimating the risk that an offender will in fact re-offend. However, it is a well-known fact that risk assessment is a notoriously controversial issue. Even though today only a few theorists would go so far as to assert that the forecasting of crime is meaningless, it is more generally accepted that there are serious challenges when it comes to the degree of reliability of risk assessments in individual cases (see e.g. Monahan and Skeem 2014, 2016; Keijser et al. forthcoming). There are well-known problems concerning both false-negatives, that is, people who are considered not to be dangerous but who nevertheless end up re-offending and, more controversially, false-positives, that is offenders who are deemed dangerous but who would in fact not have re-offended (see also Chap. 7). However, it could be held that to some extent the significance of these problems would be reduced if neurointerventions could be imposed as punishments. Consider a situation in which an offender, as the response to a criminal act, is forced to go through a neurointervention. If the offender is a false positive—someone who would not have re-offended—then the treatment might not be pointless or unjustified. It would still have served the purpose of punishing the offender for the crime that has been committed. In other words, it seems that challenges concerning the reliability of risk assessments could constitute a further reason for considering whether neurointerventions could serve punitive purposes.

Finally, there seem to be strong reasons to believe that some neurointerventions currently being used on offenders are not regarded merely as therapeutic interventions but are to some extent also imposed punitively. For instance, Forsberg and Douglas have outlined a set of possible characteristics of the use of anti-libidinal interventions on sex offenders: that such interventions are provided at the direction of correctional authorities; that they are provided only on the condition that the recipient has committed or is expected to commit a crime; that they are stopped after a fixed period of time or when an offender's risk of re-offending has decreased below a certain level; and that such interventions are described in terms of correctional language (Forsberg and Douglas 2017, p. 459). By analysing the use of anti-libidinal interventions in two jurisdictions—California and England/Wales—on the ground of such characteristics, they conclude that the intentions behind the interventions are, at least partly, to 'achieve correctional objectives' (2017, p. 453). But if this is the case, that is, if some neurointerventions are already being used in a way that is not merely therapeutic but also intended to be punitive, then it is reasonable to consider whether it is really the case that neurointerventions can, on closer scrutiny, function in this manner.

As the ensuing discussion will show, there are even further reasons for motivating a discussion as to whether neurointerventions can function as punishment. However, the outlined reasons constitute a sufficient starting point for engaging in a closer analysis of this question. In order to avoid misunderstandings, the question that will be addressed is not whether a neurointervention will always function as a punishment if it is compulsorily imposed on an offender. Obviously, this is not the case. If someone puts a drug into an offender's food and thereby affects the offender's mind, or if someone decides to perform a neurosurgical procedure on an offender who is, for other reasons, under an anaesthetic, then this will not count as the imposition of a punishment (just as it would not count as a punishment if I were to lock the door of an offender's room). Rather, the question is whether a neurointervention *if properly imposed* could function as punishment.[1]

In order to examine this question, we will start by considering whether a neurointervention can satisfy the minimal conditions set by definitions of punishment. However, since several modern theories of punishment imply that a punishment should serve several functions beyond those more narrowly figuring in the definition, it will also be considered whether neurointerventions as punishment could work within such more comprehensive theories. Having examined these questions, the focus then changes to whether it would be morally acceptable to use neurointerventions as punishment. This discussion opens up broader considerations of whether, as several theorists seem to believe, the use of neurointerventions in this way should be banned as cruel and unusual punishment, or as a punishment that violates an offender's dignity, is degrading or inhumane. It will be argued that such need not be the case. Furthermore, it is suggested that the considerations that I present not only have implications for penal theories that have been designed to take future consequences into account, but also for full-blown retributivist theories. In the end, I conclude that the initial question should be answered in the affirmative: a neurointervention can function as a punishment. However, it should be underlined what this means. The point is not to suggest that all types of neurointerventions will satisfy the requirements set by all penal theories. More modestly, the claim is that neurointerventions can be designed in such a way as to serve the function of punishment and, furthermore, that this may in some cases be desirable.

[1] I will not enter a closer discussion here of what this precisely means. The point roughly is that we can imagine that an offender is sentenced to neurointervention by the criminal court for having committed a crime.

Neurointerventions and the Definition of Punishment

In considering whether a neurointervention can be inflicted as a punishment, an obvious approach would be to specify what a punishment is and then to examine whether a neurointervention can be subsumed under this concept. However, this is not a straightforward task. A simple approach that has been suggested to this conceptual question might be to ostensibly point to all paradigmatic instances of punishment (see e.g. Montague 1995). However, clearly such an approach is not satisfactory when the question one is considering is whether a type of sanction or treatment which is not usually used as a punishment could function in a punitive manner. If we wish to examine whether a particular treatment could work as a punishment, then we are not interested in knowing whether this treatment has already been categorized as a paradigmatic instance of a punishment. Rather the point is whether it possesses the properties that are necessary and sufficient for it to function as a punishment. However, even if we turn to the discussion of a genuine definition of punishment, complications arise, namely, due to the fact that no generally agreed definition exists. Thus, what I will do here is outline a set of properties which has been suggested as constituting the necessary and sufficient conditions in the most elaborate discussions of the definition without, however, committing myself to any particular definition. The point will be to examine whether a neurointervention, if properly imposed, can satisfy even the most demanding definitions of punishment. I will consider in turn four properties crystalized during the discussion.[2]

The harm requirement. The first and most obvious property of a punishment is that it does something burdensome to the individual upon whom it is imposed. This requirement has been presented in various ways. For instance, some theorists have characterized punishment in terms of its causing pain. Others have talked of suffering or unpleasantness. However, it is obvious that a very narrow specification of precisely how a punishment is burdensome

[2] An example of a definition containing all the following requirements has been presented by Boonin who holds—what he calls a 'weak version' of a definition—that 'P's act *a* is a legal punishment of Q for offense *o* if and only if: (1) P is a legally authorized official acting in his or her official capacity and (2) P does *a* because P believes (perhaps mistakenly) that Q has committed *o* and (3) P does *a* with the intent of harming Q (even if P fails to actually harm Q) and (4) P's doing of *a* expresses official disapproval of Q for having committed *o*' (Boonin 2008, p. 25). Along the same lines, Duff and Zachary have suggested that a punishment 'involves the imposition of something that is intended to be both burdensome and reprobative, on a supposed offender for a supposed crime, by a person or body who claims the authority to do so' (Duff and Hoskins 2017).

will not be satisfactory. For instance, it is hardly appropriate to hold that a fine involves the experience of pain. Thus, a more proper way of characterizing a punishment is to say that it inflicts a harm on the individual who is punished, where harm is understood broadly as causing the individual to be worse off in some way (see e.g. Boonin 2008, p. 7). Can a compulsorily imposed neurointervention satisfy this necessary condition?

The answer must be in the affirmative. The reason we have previously said that a compulsory neurointervention is *pro tanto* wrong, is precisely that it is usually undesirable for the offender. First, the offender is compelled to go through the intervention. Second, some interventions may involve an element of suffering (e.g. a surgical procedure, even if carried out under an anaesthetic, would usually have such an effect). Third, the knowledge that one is about to be put through an intervention may in several ways be experienced as unpleasant. Fourth, a neurointervention may reduce or eliminate the possibility of engaging in activities that one regards as valuable. Thus, it seems implausible to insist that this condition cannot be met when a neurointervention is forcibly imposed.[3]

However, some theorists have underlined that, on closer scrutiny, what matters cannot be the mere fact that harm is caused to an individual. After all, there are various ways in which the state causes harm to citizens. For instance, as David Boonin has underlined, the state may charge a fee for processing a marriage licence, which is something very different from being charged a fine for having violated antipolygamy laws. And the state may place someone in pretrial detention, which appears to be something other than punishing the individual (Boonin 2008, p. 12). More precisely, there seems to be a difference between, on the one hand, doing something that foreseeably will cause harm to an individual and, on the other, causing harm as part of the purpose of what one is doing.[4] Thus, on the ground of this observation it has been suggested that what characterizes a punishment is not merely that

[3] It might perhaps be suggested that a punishment is not merely (intended to be) harmful at the time it is imposed on the offender; rather it must be (intended to be) harmful all things considered. And, further, this condition is not satisfied by a neurointervention which, even if it is (intended to be) harmful when it is imposed, may nevertheless benefit the offender in the long run. However, this objection is not convincing. First, this would imply that a prison term is not a punishment if it causes (or is intended to cause) the offender to abstain from crime in the future (after release) and, thereby, implies that the offender ends up leading a better life all things considered. Second, there may be neurointerventions which are (intended to be) harmful for an offender, all things considered, but which could nevertheless be imposed if they imply that the offender steers clear of criminality.

[4] In his early discussion of the definition of punishment, Benn strongly underlined this point when he held that the unpleasantness cannot merely be an incidental by-product of a punishment, but something that is 'essential' to punishment (Benn 1967, p. 8).

it involves harm, but more precisely that it involves *intentional* harm (Duff and Hoskins 2017). But could a forced neurointervention also satisfy this specification of the harm requirement? Is not the intention of an intervention something very different, namely, to influence the mind of the criminal?

There is no doubt that the effectuation of a neurointervention is usually born by the latter intention. However, surely this does not show that the harm caused by an intervention cannot be intended. If, as we have just seen, a neurointervention involves harm then this can be what constitutes (part of) the reason for sentencing an offender to a neurointervention. To insist that the infliction of harm cannot constitute the only purpose of imposing a neurointervention, and that it therefore cannot be a genuine punishment is not convincing. Obviously, the infliction of harm need not be the only thing that is intended when someone is punished. For a utilitarian, it may be necessary to impose harm, that is, harm may be intended, but obviously there are also other intentions involved in punishing offenders, such as crime prevention. Similarly, many modern versions of retributivism do not regard the infliction of harm as the only thing that is intended in punishing offenders. Thus, to insist that a punishment presupposes that harm is all that is intended, is not plausible. Or, as Boonin puts it in his defence of the intentional harm condition: '[t]his is not to insist, it is important to emphasize, that the offender's suffering must be intended for its own sake... Rather, it is to maintain that the punisher intends to harm the recipient of the punishment and does not merely foresee it, even if this harm is, in turn, intended for the sake of some further end' (Boonin 2008, p. 14). Thus, no matter whether one adheres to traditional definitions of punishment, emphasizing the significance of actual harm infliction, or more recent definitions focusing on intentional harm, it is clear that the idea that neurointerventions could function as punishment cannot be rejected on the ground of the harm requirement.[5]

The retribution requirement. It is generally agreed that harm (or at least intended harm) constitutes a necessary condition for something to count as a punishment. But it is equally clear that this requirement does not constitute a sufficient condition. If I kick a colleague in the leg, this may be intentionally done and harmful without being a punishment. A generally accepted further condition therefore is that a punishment involves the imposition of harm as a

[5] To avoid misunderstandings, it should be underlined that I am not holding that all neurointerventions always involve harm. There could be cases in which an offender is happy to take a drug that does not have any kind of harmful effects. In such cases, the intervention would not work as a punishment. But the same is the case with regard to other standard cases of punishment; e.g. there may be the (rare) case in which some highly desires to go to prison.

response to a transgression, that is, as a reaction to the fact that the recipient has broken the law. This is sometimes referred to as the 'retribution requirement' (e.g. Boonin 2008). Some theorists have suggested that this requirement is too strong because it conceptually blocks the possibility of punishing innocents. A possible way of dealing with this challenge has been to modify the condition by holding that a punishment is imposed on someone who is *believed* to have committed a prohibited act. However, in the present context there is no reason to consider this specification more thoroughly. As long as we are not considering the imposition of neurointerventions on persons who are merely assumed to be dangerous, but only on persons who have been found guilty of crime, it is obvious that such interventions will satisfy the requirement just as well as any other standard type of punishment.

The authorization requirement. The same is the case if we turn to another standard condition of punishment, namely the so-called authorization requirement. Even if harm is imposed on someone as a response to a transgression, this is not sufficient to turn this into an act of punishment. For instance, as Igor Primoratz puts it: 'The victim of the offense, or a relative or a friend, can take revenge on the offender, the mob can lynch him; but neither will be punishment' (Primoratz 1990, p. 4). The point is that a punishment must be carried out by a judge, an executioner, or at least some party who is authorized to do so by the legal order against which the offence has been committed. Turning to compulsory neurointerventions, it is obvious that this requirement can be satisfied. Even though the imposition of a neurointervention will involve physicians or other healthcare personnel, who do not constitute the standard executioners of a punishment, it is clear that they could be acting with proper authorization from the legal system. This is already the case when doctors take part in different types of treatment and some types of punishment (say, the death penalty). Whether physicians should be involved in the execution of punishments is of course another question to which we will turn in the following chapter.

The reprobation requirement. The final necessary condition for something to constitute an instance of punishment has been referred to as the reprobation requirement. This requirement was often ignored in the discussion of punishment that took place in the middle of the last century. However, in an influential article, Feinberg emphasized this deficiency in standard definitions of punishment, namely, that they had generally ignored the fact that punishment, in contrast to mere penalties, serves as a device for the 'expression of attitudes of resentment and indignation, and of judgments of disapproval and reprobation' (Feinberg 1965, p. 98). In other words, rather than being a harm simpliciter, punishment has, as Feinberg put it 'symbolic significance'. This idea has subsequently been developed by many theorists who do not merely perceive the expressive element of punishment in relation to the discussion of a proper definition, but see it as the *raison d'être* of punishment, that is, as playing a crucial role in the justification of punishment. Some of the most influential modern versions of retributivism are based on the view that a punishment serves as a sort of communication in which an appropriate condemnatory message is conveyed to the offender. Can the imposition of a neurointervention on an offender serve this function?

From the outset, it is hard to see why this should not be the case. Theorists who have stressed the significance of the reprobation requirement and, more generally, adherents of expressionist versions of retributivism, have in my view often not been very precise when it comes to the question as to how the communication involved in a punishment takes place. It has frequently been underlined that there seems to be a difference between a fine and an extra tax, or between imprisonment and a quarantine. The former kinds of treatment deliver a condemnatory message, while the latter do not. However, how precisely the communication takes place—e.g. when and how you receive the conveyed message when you spend time in prison—is not very clear. In other words, the reprobation requirement has not been supplied with a semantical theory that shows precisely how the meaningful messages conveyed in the punishment of someone, are generated.[6] But in the absence of such a theory, it is very hard to see why a compulsory neurointervention should not be able to serve a communitive purpose. Recall again, that we are not merely considering an intervention that is forced upon someone. Rather what is at stake here is that a perpetrator

[6] It might be suggested that a punishment conveys a message by being a *conventional sign* of reprobation. For instance, Feinberg holds that in the same way as 'certain words have become the conventional vehicles in our language for the expression of certain attitudes . . . champagne is the alcoholic beverage traditionally used in celebration of great events . . . black is the colour of mourning' . . . a punishment constitutes a 'conventional device' for the expression of certain attitudes towards the offender (Feinberg 1965, p. 402). Furthermore, it could be held that this is precisely the reason a neurointervention cannot convey appropriate messages: such interventions are usually regarded as treatments and do not constitute conventional signs of reprobation. However, this strikes me as premature. The crucial question obviously is: When will a neurointervention be perceived as merely a treatment and when will it be seen as a punitive reaction? If an offender is sentenced to a compulsory treatment for having committed a crime, then it seems reasonable to believe that this conventional framing is sufficient to give the treatment a punitive content and to convey reprobation (at least to the same extent as other types of harmful sanction meted out by the criminal court). Surely, an offender will perceive a neurointervention imposed in this manner as something very different from a standard treatment in healthcare practice.

is sentenced to a neurointervention by the criminal court and that, as we have seen, such an intervention is experienced as harmful by the offender. If this is the case, then it is very difficult to see why this type of sentence must be devoid of meaning; that is, why it cannot possibly serve as a method of delivering an appropriate condemnatory message. Indeed, it is hard to believe that a semantical theory can be developed establishing that other types of hard treatment, to which an offender is sentenced, can convey messages, while the harm involved in a neurointervention to which an offender is sentenced cannot possibly do so.

However, it might perhaps be objected that this answer is premature. Perhaps what is at stake is not that a properly imposed neurointervention cannot possibly serve a communicative function. Rather the point is that it cannot convey the right kind of message. More precisely, it might be held that, if the harm involved in a forced neurointervention can convey a reprobative message, this will nevertheless be blurred by the fact that the offender also gets the message that he is sentenced to this particular type of treatment with the purpose of causing a change in his criminal tendencies. In brief, it might be suggested that a neurointervention is not a priori excluded as a means of communication, but rather that it will send 'mixed signals' and thereby fail when it comes to the delivery of a genuine reprobative message.

However, this view seems very hard to sustain. As we have already seen, the infliction of harm need not be the only thing that is intended for something to work as a punishment. In fact, there will often be other purposes related to the imposition of a punishment. And the fact that these purposes are recognized by the offender will not rule out the possibility that the offender is also properly condemned. It is in this connection worth recalling that most modern penal theorists (including retributivists in the expressionist camp) believe that there are various kinds of intermediate sanction that can serve as alternative punishments, and not merely as alternatives to punishment. But the fact that an offender realizes that he is sentenced to a period of home detention or community service, rather than a brief period in prison, because these sanctions are preferable in terms of resocialization or because this will save the state some money, does not imply that such sanctions do not constitute genuine instances of punishment sending a reprobative message. In brief, if one accepts that a punishment can be inflicted for several purposes, then it is not convincing to maintain that the different purposes of a properly imposed neurointervention will block for the possibility that the offender gets the message that he is being blamed for his misdeed. Thus, it is very difficult to see that a neurointervention cannot be imposed in a manner that will satisfy the reprobation requirement.

In summary, what we have seen is that, given what has standardly been regarded as the set of necessary and jointly sufficient conditions for something to serve as a punishment, it is not plausible to contend that a neurointervention cannot serve as a punishment.[7] A compulsory neurointervention can involve harm, it can be imposed as a response to criminal conduct, it can be carried out by someone who works under the proper authorization, and there is no reason to believe that it cannot send a proper message to the offender. However, even if all this is correct, it may not be sufficient to convince all penal theorists. As we shall now see, some theorists have emphasized that a punishment must serve purposes that go beyond the more generally accepted definitional requirements.

Further Penal Theoretical Requirements

The considerations in the previous section are not merely conceptual. Even though the purpose has been to investigate whether a properly imposed neurointervention can satisfy the conditions outlined in definitions of punishment, the point of doing so is of course not purely linguistic. If we were only interested in whether a neurointervention falls under the extension of the concept of punishment, then there would be nothing in this discussion that prevents the imposition of neurointerventions as a reaction to criminal conduct, as long as we call this response something other than punishment (say neurishment). The point of the investigation obviously is that a punishment is believed, at least according to most theories, to constitute a morally proper reaction to crime. However, it is also clear that, on closer inspection, many of the various ethical theories of punishment currently dominating the field operate with more comprehensive views on what a punishment should achieve. In other words, according to such views, in order for a punishment to be justified it will have to live up to more requirements than the

[7] It might be objected that the fact that each of the necessary conditions can be satisfied by some neurointerventions does not establish that there are some neurointerventions that satisfy *all* the necessary conditions (which, of course, is crucial in order for an intervention to constitute a punishment). However, this objection is hard to sustain. As indicated, most compulsory neurointerventions will be harmful in one way or another. Arguably, all such interventions can be imposed as a reaction to criminal conduct and can be imposed with the proper authorization. Thus, if a least a subgroup of the interventions can satisfy the reprobation requirement, then it seems at least this subgroup of interventions will satisfy all the necessary conditions.

minimal conditions specified in standard definitions of punishment. This means that adherents of such theories may admit that properly imposed neurointerventions can be characterized as a punishment while, at the same time, holding that such interventions constitute morally unacceptable types of punishment because they fail to live up to more ambitious accounts of what a justified punishment is supposed to achieve. The purpose in this section, therefore, is to outline some of the extra conditions that have been suggested as crucial for a punishment to be justified, in order to examine whether more comprehensive accounts exclude the possibility that neurointerventions can serve a proper punitive function.

What this means is that the discussion will be more complicated than the definitional considerations in the previous section. A rejection of a neurointervention as a proper type of punishment will be premised, both on the facts that the more comprehensive theory of punishment is plausible and that the theory implies that neurointerventions cannot satisfy these requirements. However, in the present context I cannot engage in a sufficient discussion of the first sort of premise. Basically, this would amount to a discussion of what constitutes the most plausible theory of punishment which, obviously, is a question that cannot even tentatively be answered here. Thus, what I will do instead is to focus on the second premise; that is, it will be considered whether the requirements that have been suggested manage to establish that neurointerventions cannot function in a proper punitive manner. It will be argued that, even though this may well be the case for some types of neurointervention, there are also other types where the answer seems to be the opposite. In particular, it will be suggested that neurointerventions that involve cognitive enhancements in offenders can live up to many of the requirements.

The non-exclusion requirement. A possible way of assessing ethical theories of punishment as well as tendencies in modern penal policy, is in terms of the distinction between inclusionary and exclusionary treatment. Some theorists have underlined that a plausible theory of punishment must be characterized by upholding a non-exclusionary approach to offenders. For instance, Nicola Lacey holds that '[n]ormatively . . . we might expect liberal-democratic criminal justice to aspire to be reintergrative and inclusionary rather than stigmatizing and exclusionary'. However, in her view this expectation forms the background for what she refers to as 'one of the most troubling empirical paradoxes of contemporary democratic justice', namely, the fact that 'in many countries, criminal justice policy has been driven in an exclusionary direction with—perhaps even because of—popular, and hence literally democratic, support' (Lacey 2007, pp. 2–3). As has been underlined by other theorists, exclusionary treatment of citizens can take various forms. For instance, Antony R. Duff distinguishes between political exclusion, by which citizens are deprived of the possibility of taking part in the political life; material exclusion, which happens when citizens are denied a fair access to societal resources and benefits; normative exclusion, which concerns the extent to which citizens are treated as if societal values do not apply to them; and linguistic exclusion by which citizens are addressed in a language they do not understand by other citizens or by the state (Duff 2001, pp. 75–77). Focusing more narrowly on sentencing, it is clear that there are different ways of dealing with offenders which instantiate one or several of these ways of exercising exclusion. For instance, disenfranchisement constitutes an obvious example of a way of excluding offenders from political participation. Another example could be a punishment which only serves the purpose of incapacitating an offender. Duff even believes that a system that imposes punishment with the sole purpose of rational deterrence, should also be regarded as exclusionary (Duff 2001, p. 78). But what should we say about forced neurointerventions? If it is plausible to contend that an ethically acceptable type of punishment should treat offenders in an inclusionary manner (or at least not exclusionarily), does this rule out the use of neurointerventions as a response to criminal conduct?

In order to answer this question, it is important to underline once again what we are here considering. The point is not to examine whether neurointerventions *can* be used in an exclusionary manner. There is no doubt that this is the case. For instance, as mentioned in the earlier discussion of the identity objection in Chapter 3, a radical instance of the manipulation of a person's brain could be one that results in a genuine shift of numerical identity. And, as would normatively amount to the same, the killing of a person can also be regarded as a way of intruding into the brain and mind of a person. It is hard to dispute that such ways of influencing an offender cannot, in any plausible interpretation, satisfy a requirement of non-exclusion. Moreover, it may be the case that even less extreme examples of neurointervention can sometimes contribute to societal tendencies of stigmatizing offenders. Whether there is a risk that neurointerventions will work in this manner in the current political system, which is characterized by the exclusionary dynamics to which Lacey and others have rightly directed attention, is a question to which we will turn in a later chapter (Chap. 7). However, the question we are discussing here is whether forced neurointerventions can in principle live up to what is required from a proper type of punishment. And it is hard to see why this should not be possible. If an offender is sentenced to some sort of neurointervention, and if this is done with the intention of reducing the risk that the offender will engage in future criminal conduct, then this practice will be effectuated for some of the same purposes as more traditional methods of rehabilitation. As we have seen, and as we will return to shortly, if such treatments work as they should, they could even be used as alternatives to other types of punishment which from the outset are of a much more exclusionary nature (such as imprisonment or home detention). Thus, given the fact that a neurointervention can be imposed as a method of rehabilitation and reintegration, it is reasonable to conclude that it can also live up to a requirement of non-exclusion.

Some critics will no doubt object that, even if it is true that a forced neurointervention, if properly imposed, does not constitute a way of getting rid of an offender, but rather a way of maintaining the offender as a member of the society, this is not sufficient to fully satisfy the requirement. One could adhere to much more demanding theories of what it takes to treat an offender in a genuinely inclusive manner. For instance, Duff has developed such an account (Duff 1986, 2001).[8] However, we will return to some of the more detailed requirements of such a theory below. For the present it is reasonable to hold that a neurointervention is not properly categorized as a way of separating the offender from the community.

The attention requirement. Another condition stressed by recent penal theorists, is that an important aspect of what a proper punishment does is to help in focusing the offender's mind on the wrongdoing that has been committed. This requirement can be given slightly different interpretations. One possibility is to hold that a punishment should provide a framework that makes it possible for the offender to focus his attention on the misdeed. For instance, Elizabeth Shaw has suggested that incarceration may serve this function. What she says is that 'prison may be justified, because it allows the offender to dwell on the seriousness of his crime without being distracted by the competing claims on his attention that he would experience if he was still living in the community, e.g. employment, socializing and family obligations' (Shaw 2018, p. 260). Another interpretation would be to suggest that a

[8] Interestingly, Duff has defended the view that an exclusionary approach may be adopted against some persistent offenders (see Duff 1998). In his more recent work, he seems less convinced as to what constitutes the most plausible approach to genuinely persistent offenders (Duff 2001).

punishment should not only provide a background for attention by eliminating factors that might distract the offender, rather it should more actively assist in directing the offender's mind to the nature of the crime. For instance, Duff has for this reason defended the use of community service orders as a desirable type of punishment. In a case where an offender has committed some sort of vandalism and is subsequently forced to work at repairing the effects, such an order may bring home to the offender the nature and implications of the crime. As he puts it: 'By confronting her with the damage done by vandalism and the work involved in making it good, it says to her in effect "Look what you have done!"' (Duff 2001, p. 105). Obviously, one need not hold that an offender's focus on the crime is desirable in itself. Attention on a crime may well be seen as the first step in a process that serves further ends. What these could consist in is something to which we will return shortly. However, the important thing is that a proper punishment should provide a framework for directing the offender's mind onto the crime and its consequences either by removing distracting factors or, if possible, by reminding the offender of the nature of the crime. Can a compulsory neurointervention possibly live up to this requirement?

From the outset it seems that the answer must be in the affirmative. In so far as a neurointervention of some sort can influence and improve the cognitive capacities of an offender, it would seem that it can also contribute to concentrating the offender's mind on his wrongdoing. For instance, suppose that an offender suffers from an attention-deficit disorder such as ADHD—which, as earlier noted, is highly prevalent in prison inmates (Ginsberg and Lindefors 2012)—in that case a proper medical treatment (such as standard methylphenidate treatment) may contribute to increasing the offender's power of attention. As a more controversial example, evidence has been presented to the effect that psychopathic behaviour that threatens social conformity is not—as has been assumed—a result of an amygdala-mediated deficit in fearlessness but rather attributable to an attention-related deficit that undermines the processing of information. On the grounds of a study of 125 psychopathic and non-psychopathic prisoners, it was concluded that '[a]lthough psychopathic individuals are commonly described as cold-blooded predators that are unmotivated to change, the present findings identify an attentional dysfunction that may account for their chronic disinhibition and insensitive interpersonal style and provide a meaningful target for early clinical intervention' (Newman at al. 2010, p. 70). If these results can stand up to further scrutiny, and if it is the case that the attention deficits in this group of offenders impede the possibility of directing appropriate attention to the crime that has been committed, then it seems that a neurointervention that would remedy this deficit could be helpful if seen in the light of the attention requirement. Similarly, even if we consider a group of offenders not suffering from any pathological disorders, it might still be possible that a neurointervention could either improve the relevant attention capacities or reduce the influence of attention-distracting factors.

However, even if all this is accurate, it might be objected that these considerations are somewhat misplaced. After all, what we are considering is not whether a neurointervention can possibly serve as a remedy when people are suffering from attention deficits, but whether a neurointervention can function as a punishment. In other words, it might be objected that, while imprisonment or a community service order can both be harmful to the offender and of help in directing the attention of the offender to the crime that has been committed, this is not the case with regard to a neurointervention. And the reason simply is that the harmfulness involved in the neurointervention does not provide the proper background for focusing on previous misdeeds. For instance, this seems to be the view held by Shaw. Even though she underlines the possibility that neurointervention can increase the power of an offender's attention, she also rejects that such an intervention can work as a punishment. What she holds is that, if a neurointervention satisfies the harm requirement of a punishment, for instance, by the harmful side-effects of a treatment, such as nausea or headaches, then this type of harm is 'more likely to interfere with the offender's capacity to engage in moral reasoning about the offence' (Shaw forthcoming, p. 16). Now, this seems absolutely right. Few people have any power left for attention in a state of nausea. However, the answer obviously is that the harm requirement can be satisfied in other ways than by induced nausea. For instance, as we have seen, it could be held that the harm consists in the fact that the neurointervention is forced upon the offender against his or her will. In this respect the neurointervention need not be different from a community service disorder, which may be harmful mainly because it is non-consensually forced upon the offender and not necessarily because the activities it involves are harmful in themselves. Furthermore, there is nothing that prevents a neurointervention being imposed in combination with other sorts of harmful treatment. For instance, it could be the case that a prison term (or a community service order) is combined with the imposition of a neurointervention thereby constituting an overall punishment and one that would be more harmful, and hence more severe, than if the prison term was not accompanied by the compulsory intervention.[9] In such a case, the possibility of the offender directing attention to the crime may be even better than if the prison term was served without the neurointervention. In more general terms, it should once again be kept in mind that what we are considering here is not whether all types of neurointervention (including those that may cause nausea or headaches) can serve a punitive function, but rather whether a neurointervention can be so designed as to make it possible for it to function in this way. As argued, it is hard to see that an attention requirement blocks this possibility.

The remorse requirement. Whether an individual who has committed a crime subsequently reacts to his own misconduct in a remorseful way is something that plays a significant role in penal practice. It is a well-settled legal principle in many jurisdictions to treat an offender's remorse as a mitigating factor in sentencing (see e.g. Tudor 2008). For instance, surveys have shown remorse to constitute the most commonly invoked mitigating factor for several types of offence in England and Wales (Pugh and Maslen 2017, p. 500). Moreover, theorists have held that the significance of remorseful reactions in offenders is theoretically well-founded. For instance, the view that remorse *should* serve as a mitigating factor has been defended by Steven Tudor, who even suggests that the force of a theory of sentencing partly depends upon the extent to which it succeeds in satisfactorily accounting for the mitigating role of remorse (Tudor 2008, p. 242). However, even if one disagrees on the principle that remorseful offenders should be treated more leniently than offenders who do not display any remorse, one might well adhere to the more general view that it is morally desirable that an offender is sincerely remorseful and, therefore, that a proper type of punishment is

[9] It might be objected that if one combines different types of sanction, the result will not be one overall punishment by rather several separate sanctions. However, this contention is not persuasive. In many jurisdictions, it is already the case that offenders can receive sentences involving different types of sanction. For instance, an offender can receive a period of incarceration and a fine. Moreover, it is possible to imagine that two different types of sanction can be more narrowly integrated; for instance, a neurointervention could remove an attention deficit and another type of sanction could seek to mimic the nature of the crime committed. In this way, the two sanctions will complement each other in directing the offender's attention to his or her crime. I see no reason why such a reaction could not be imposed as one overall punitive response to the crime committed (and be apprehended in this way by the offender). Finally, even if one believes the attention requirement to be crucial, there is no reason to hold that the requirement must be satisfied by each of the separate components of a sentence. In other words, even if one insists that a combined punishment will not constitute one overall punishment, but two separate punishments, the imposition of a neurointervention could still satisfy the requirement as long as it is combined with another type of punishment that satisfies the requirement.

one that may succeed in facilitating remorse in offenders (see e.g. Duff 2001). Can a punishment *qua* a neurointervention satisfy such a requirement?

Suppose that what we mean by remorse is roughly the emotional suffering experienced by a person who recognizes that he or she acted wrongly against another person. In that case it might seem that the question we are now considering is whether it is possible to use neurointerventions as a direct way of inducing the relevant sort of emotional suffering in offenders. However, even if this question can be answered in the affirmative—it will probably be possible in various ways to cause emotional suffering by the use of neurointerventions—it is also clear that this way of framing the question is premature. What one is interested in, if one regards remorse as a morally desirable reaction to crime, is not merely that an offender is suffering but that he is suffering for the right reasons and perhaps with the right motivational implications. As Jonathan Pugh and Hannah Maslen have recently argued, remorse is a complex psychological phenomenon that contains several interrelated elements (Pugh and Maslen 2017).

First, there is a cognitive side to remorse. Briefly put, an offender must be able to recall and understand the nature of the wrongful act. Second, there is an affective element in the obvious sense that an offender must experience some sort of aversive affect. Third, remorse also involves a motivational component. A genuinely remorseful offender will be motivated not to repeat the wrongful act, to apologize to the victim, and to offer some sort of reparation for the wrongdoing. Furthermore, beyond the fact that remorse has cognitive, affective, and motivational sides, it is also clear that these elements must be properly connected. If an offender understands that he has acted wrongly but experiences aversive affects for other reasons, and is motivated to repair the loss of the victim solely out of reasons of self-interest, then though all three elements may be present, they will not form an instance of genuine remorse because the reactions are not properly related. This shows why a technologically induced aversive emotional reaction in an offender may not constitute an instance of remorse, even if one imagines that the affect per se is identical to the affect found in genuinely remorseful offenders. As Pugh and Maslen note: 'The wrong that one committed is the central object of remorse; the aversive affect does not, so to speak float free from one's belief about one's culpability for the wrong' (Pugh and Maslen 2017, p. 512).

Interestingly, however, Pugh and Maslen's tripartite account of remorse also shows how it can be possible for a neurointervention to facilitate *genuine* remorse. If reduced cognitive capacities somehow hinder the possibility for an offender to remember or fully understand the nature of a crime, then a neurointervention that would remedy these cognitive impediments may constitute the first step in a process the could prompt subsequent and proper affective and motivational reactions in the offender. In short, neurotechnological improvement of cognition may constitute a way of facilitating genuine remorse. As an example, Pugh and Maslen mention the use of neurotechnology as a means of facilitating memory retrieval. Several studies have indicated that there may be various ways of improving memory recall pharmacologically (e.g. Grön et al. 2005; Croxson et al. 2012). Other studies have indicated that episodic memory can be modulated by the use of neurotechnological means such as transcranial direct current stimulation (Manenti et al. 2013). Moreover, it has been suggested that pharmacological methods could be used as a way of facilitating remorse by increasing the understanding of the nature of a crime in offenders suffering from empathy deficits. Presumably, as we have seen above, such understanding could also be improved pharmacologically in some offenders suffering from attention deficits.

Therefore, it is not plausible to reject neurointerventions as a type of punishment on the grounds of the remorse requirement. As long as it is kept in mind that it cannot reasonably be required that a punishment must succeed in generating a remorseful reaction in offenders—this requirement would rule out all types of punishment, since such a reaction can never be guaranteed—but only that it should serve a *facilitating* function, there is no reason on this ground to reject that neurointerventions can be imposed in a punitive manner. Noteworthily, the same overall point can be made more broadly in relation to penal theories that do not focus specifically on remorse but which regard it as crucial that a punishment should clear the ground for other types of response in offenders. For instance, it might be suggested that a punishment should function as a vehicle for some sort of self-reformative process; that is, not a kind of reform that is directly induced in the criminal but one that the offender himself is initiating and working himself through (Duff 1986, 2001). Alternatively, it could be suggested that a punishment should generate other types of affective reaction than remorse, whatever these precisely may consist in. However, as long as all such reactions should not be guaranteed—which, as just noted, would constitute an implausible condition—but rather be something that a punishment should facilitate, then it seems that neurointerventions operating at a cognitive level, might well provide the preconditions on the grounds of which such further reactions can rise. Thus, neither the remorse requirement nor requirements of other cognate reactions in offenders seem to block the possibility of a punitive use of neurointerventions.

The rationality requirement. The significance of the fact that people are addressed as rational and responsible beings has, for understandable reasons, been underlined in discussions of various moral issues. For instance, in general considerations of manipulation, or what she refers to as 'non-argumentative influence', Jennifer Blumenthal-Barby contends that: 'Non-argumentative influence is influence that operates either by bypassing a person's awareness or by relying on facts about the subject's psychology, such as knowledge about his emotions, how he perceives things, how he makes judgments and decisions, and what he desires. It is in contrast to influence that operates by engaging with a person's rational capacities and offering him reasons and arguments (i.e. rational persuasion)' (Blumenthal-Barby 2014, p. 123). This distinction between ways in which another person is addressed has, by several penal theorists, also been underlined as crucial in the distinction between acceptable and unacceptable punitive reactions to crime. A proper punishment must treat the offender as a rational being and, accordingly, address him by engaging with his rational capacities. But what does this mean? What precisely does it take to satisfy this requirement? As we will see, the requirement can be given different interpretations and, as will be argued, it is hard to see that any of these interpretations can be used as a background for rejecting that neurointerventions can serve a proper punitive function.[10]

A first interpretation, and one that can easily be dismissed, would be to contend that all the state should do is address offenders as rational beings by, literally speaking, engaging in a dialogue in which reasons and arguments are presented to the offender in precisely the same way as when two interlocutors are engaged in a normal rational conversation. However, this obviously cannot be a requirement for a plausible theory of *punishment*. A punishment is something more or different from simply presenting an argument to

[10] It should be mentioned that, if a compulsory neurointervention is imposed on an offender as a type of punishment, this obviously does not imply that the method and purpose of the intervention should not be explained to the offender. In fact, it would be absurd not to provide the offender with a proper explanation. The fact that an intervention is imposed *because* the offender has committed the crime does not preclude explaining why this type of punitive reaction is regarded as preferable. Analogously, a sentencing judge can also tell an offender why, in his case, a period of home detention is regarded as preferable to a period of imprisonment. Thus, the requirement that an offender should be treated as a rational being must mean something beyond the fact that the offender should be addressed with this sort of explanation.

another part. Thus, this requirement would commit one to a rather extreme version of abolitionism to which no one, to my knowledge, adheres.

A much more obvious interpretation would be to hold that the sanction that should be imposed on an offender must engage with the offender's rational capacities by conveying to this person an appropriate condemnatory message. This, as we have seen, is precisely what adherents of expressionist (or rather communicative) versions of retributivism hold. For instance, as Andreas von Hirsch has illustratively put it, if one imagines a situation in which the state inflicted only a *neutral* sanction on offenders, that is, a sanction that did not convey disapproval, then this would treat offenders 'as tigers might be treated in a circus, as beings that have to be restrained, intimidated, or conditioned into compliance because they are incapable of understanding why biting people (or other tigers) is wrong' (von Hirsch 1993, p. 11) Conversely, a genuine criminal sanction conveying condemnation treats the offender as someone capable of choice and understanding.

However, if this is what it takes to satisfy the rationality requirement, then strictly speaking the requirement does not add anything that goes beyond what is already included in the definition of punishment (which, as we have seen, usually contains a reprobation requirement). And, more importantly, if, as argued above, it is correct that a properly imposed neurointervention can—as a type of harm inflicted in response to wrongdoing—serve a communicative function, then it seems that such an intervention will satisfy the rationality requirement. A properly imposed neurointervention can, as in other types of punishment, send an appropriate message to an offender.[11] However, it might perhaps at this point be felt that something is going wrong. If the state should treat offenders as rational beings when they are being punished then, even though a forced neurointervention can serve as a way of conveying a message to the offender, is it not the case that such interventions, if they at the same time seek to reform the offender, will somehow fail to fully treat him or her as a rational being? In short, can one treat someone as a rational being while at the same time attempting to change this person? If, as earlier argued, it is not the case that the content of the condemnatory message conveyed by a properly imposed neurointervention is somehow undermined

[11] It might perhaps be objected, that the fact that a properly imposed neurointervention can convey a condemnatory message to an offender is not sufficient to satisfy the rationality requirement in this interpretation. This will also require that the message reflects the gravity of the crime that has been committed. However, as we will see shortly, there is no reason to believe that neurointerventions cannot satisfy a requirement of proportionality.

by the fact that the intervention is also designed to cause alterations in the individual, then it seems that such an intervention can satisfy the standard view of what it means to address offenders in a condemnatory (and hence communicative) manner. However, as we shall now see, one could hold that the worry reflects a stricter interpretation of what it means to hold that a proper punishment should engage with the rational capacities of an offender.

It might be suggested that the requirement is not satisfied merely by engaging in a communicative interaction with the offender. A punishment must be capable of more. Ultimately, it should lead to a reform of the offender; however, and this is crucial, it should do so in a particular way. The theorist who has presented the most comprehensive and ambitious account of this idea is Duff. In his view, a punishment should ideally initiate a kind of self-reform. A punishment should not forcibly cause an alteration in an offender. Rather it should involve the use of compulsion as a way of focusing the offender's mind on the crime that has been committed and thereby provide a basis for what is ultimately the offender's own decision to engage in a process of reform. In Duff's words, a proper punishment forces an offender 'to hear' the message that the punishment aims to communicate; it should not force the offender to be 'persuaded by it' (Duff 2001, p. 126). But if this is how the rationality requirement is interpreted, that is, as a way of forcing attention without itself causing a reform in the offender, can a properly imposed neurointervention then satisfy the requirement?

Once again, I believe the answer to be in the affirmative. As we have already seen in the above discussion of the remorse requirement, a neurointervention that contributes to improving an offender's cognitive capacities may constitute a way of forcibly directing the offender's attention to the crime that has been committed and may in this way force the offender 'to hear' the message, while at the same time leaving it open for the offender to ultimately decide to engage in whatever a self-induced process of reform requires. As with other types of punishment, a neurointervention—though hopefully even more effectively—could function as a tangible way of facilitating attention.

However, even if this is correct, there may still be some who would object that, as long as *some* sort of alteration is forced upon the offender's mind, this is sufficient to violate a rationality requirement.[12] After all, if an offender's attention is improved by an neurointervention, this is the same as causing an

[12] Shaw has presented on objection along this line in the following way: 'If the state were forcibly to 'trespass' into the offender's mental realm and alter his manner of reasoning (by making him more alert and focused) this would arguably blur the boundary between the two parties in the dialogue.

alteration in the offender which in itself is not rationally mediated. In short, it might be held that the requirement is violated by the fact that an alteration operates directly on the mind and that this is so independently of what kind of alteration we are considering (i.e. even if it involves some sort of cognitive improvement). However, if this is what one holds, then the rationality requirement basically collapses into the kind of mental non-interference objection that was discussed and rejected in the previous chapter.[13] Therefore, if we leave this interpretation out of consideration, along with the radical abolitionist view that the state should do nothing beyond literally confronting the offender with arguments and reasons, it seems that the rationality requirement interpreted either as the view that a punishment should convey an appropriate message to the offender or that it should coerce the offender into 'hearing' the message, can be satisfied by some properly imposed neurointerventions.

The proportionality requirement. A final requirement that might come to mind concerns the possibility of maintaining a principled distribution of punishment. Even though there exist significant differences between the various versions of retributivist theories that have dominated the modern area of penal theory, it has been generally accepted that the principle of proportionality constitutes a plausible principle governing penal distribution (see e.g. Davis 1992; Duff 2001; Kleinig 1973; Moore 1997; Scheid 1997; von Hirsch 1993, 2017). The severity of a punishment should reflect the seriousness of the crime. This principle has been regarded as crucial for positive

Dialogue requires two independent parties—two clearly separate minds. If one party not only provides the other party with reasons to engage with, but also directly controls, through brain modification, the manner in which he engages with those reasons (albeit not the end result of the reasoning process), arguably this does not count as genuine dialogue' (Shaw, forthcoming, p. 19). However, it is left unclear why one should accept this view. Suppose that I have cheated a very excited person into taking a pill that will calm him down, and that I have done so in order to get a clearer picture of what this person believes on an important matter (and under the assumption that this person's capacity for reasoning is much better if he is not in a state of excitement). Why should we hold that I cannot, under these conditions, have a genuine dialogue with this person? Or what if I had simply asked the excited person to have a seat, knowing that this will reduce the excitement and thereby improve his reasoning capacity? Would this a priori exclude the possibility of having a genuine dialogue? I must admit that I find it hard to believe that this would be the case. Of course, one might still hold that it would be unacceptable to cheat the person into taking the pill in the previous example. But if so, then this has to do with reasons other than what is required of a genuine dialogue (e.g. that one should not cheat others—which is not relevant when we are talking forced neurointerventions—or that interference in another person's mind is simply wrong—in which case the objection seems to collapse into the previously discussed objection regarding mental non-interference).

[13] The view that any kind of alteration in the criminal should be rationally mediated, would probably not be satisfied by any sort of punishment. For instance, as we have seen, imprisonment may have various sorts of mental effect on prisoners which cannot plausibly be held to be rationally mediated.

retributivist theories according to which an offender should receive no more and no less than the punishment that is proportionate to the gravity of the crime. But the principle also plays a significant role in most versions of mixed theories of punishment; these often invoke proportionality considerations in the determination of upper constraints on what constitutes an acceptable punishment for a crime of a certain degree of seriousness. Therefore, even if a forced neurointervention can satisfy the minimal definitional conditions for what characterize a punishment, it could still be held that such an intervention should be rejected as a part of penal practice if it cannot be distributed in accordance with the proportionality principle.

However, it is easy to see that this requirement should not constitute a genuine worry. What it takes for a punishment to be distributable in a proportionate manner is that it is comparable in terms of severity to other punishments. This means that ordinal proportionality can be observed if it is possible to compare different neurointerventions with regard to their relative degree of severity and, correspondingly, if it is possible to make the same sort of comparison between neurointerventions and other standard types of punishment. However, even though what basically constitutes the proper yardstick for the measurement of severity is controversial (see e.g. Kolber 2009; Ryberg 2004, 2010; von Hirsch and Ashworth 2005), it is hard to dispute that some interventions will affect the life of an offender more than others. A forcible imposition of a drug with a transient effect will be less harmful than a forced neurosurgical procedure with long-term side-effects. Correspondingly, there is no reason to suggest that forced neurointerventions should not be comparable to other types of punishment. In the same way as proportionalists believe that different types of punishment—say fines, probation, and imprisonment—are comparable in terms of severity, it is also clear that different types of neurointervention are comparable to these punishments, for instance, in terms of the harm they cause to the offender. What this means is obviously not that there are no theoretical obstacles when it comes to the sort of ranking of punishments in severity which a proportionalist penal scheme ultimately requires (in fact, I believe there are many challenges).[14] But these obstacles have nothing in particular to do with the relative comparison of neurointerventions with

[14] For instance, the challenges consist in the facts that one and the same punishment may, for various reasons, affect offenders very differently and that it may not be a simple task to specify what part the harm that a punishment causes should be regarded as punitive harm (see Ryberg 2004). However, these challenges arise for all types of punishment.

standard types of punishment. Therefore, the answer to those who are concerned about the proportionality requirement is that there no reason to believe that forced neurointerventions will in this respect fare any worse than other types of punishment.

Then, where do all these considerations of whether compulsory neurointerventions can satisfy the various outlined requirements lead us? The answer, I believe, is straightforward. What we have seen is that, even when it comes to many of the requirements that go beyond those set in standard definitions of punishment, but which have nevertheless been emphasized as crucial for a sanction to work as a genuine punishment, there is no grounds for holding that these requirements cannot be satisfied by *some* types of properly imposed compulsory neurointerventions. Obviously, this is not tantamount to holding that adherents of penal theories that invoke any of these requirements are committed to accepting the use of neurointerventions as punishment. As we will now see, there could be other reasons for rejecting the imposition of neurointerventions as a response to crime even if one admits that they can in principle function in a punitive manner.

Punishments That Should Be Banned in Principle

What has been considered in the previous section is a number of requirements which are not usually regarded as belonging to the set of conditions that are necessary for something to count as a punishment but which have been presented as conditions that must be met in order for a punishment to be justified. That is, they do not constitute definitional requirements but additional ethical requirements that must be satisfied by a morally proper punishment. However, as just indicated, even if it is the case that a properly imposed neurointervention can function as a punishment and, furthermore, if such an intervention can sometimes even satisfy the previously considered extra moral requirements, this does not per se establish that the intervention constitutes an acceptable type of punishment. It might well be held that there are other requirements concerning characteristics which a punishment must *not* possess. In short, it might be held that there are some types of punishment which should never be imposed on offenders.

The idea that there exist such punishments that should be banned in principle has been defended by many penal theorists. For instance, even Kant—who in his exposition of *lex talionis* recommends the use of various types of punishment which modern penal theorists would regard as downright unacceptable—held that a punishment should 'be kept entirely free from any maltreatment that would make an abomination of the humanity residing in the person suffering it' (Kant 1965, p. 132). In more recent, mainly retributively-minded, penal theory, the idea has usually been presented as a rejection of punishments which are cruel and unusual, inhumane, degrading or dignity-violating (see e.g. Duff 2005; Kleinig 1998; Murphy 1979; Ryberg 2004, 2018). Thus, the question we are faced with now is whether a punishment *qua* a neurointervention belongs in these categories of unacceptable punitive reactions. A simple glance at the literature reveals that several theorists believe this to be the case. For instance, John F. Stinneford has strongly criticized the current use of chemical castration by some states in the US, holding that this practice robs offenders of rights that are integral to 'human dignity' (Stinneford 2006, p. 596). Richard G. Boire holds the compulsory use of pharmacotherapy on prisoners to be cruel and unusual (Boire 2010, p. 287). And, in the same vein, Bomann-Larsen contends that the use of central nervous system interventions as a form of punishment would not only require a total reconsidering of the penal system but also 'a setting aside of the fundamental principle that punishment should not be "cruel and unusual"' (Bomann-Larsen 2013, p. 76). Other theorists have expressed similar views. Thus, there are good reasons to subject such views to a closer scrutiny. However, before doing so, two comments must be made.

First, the view that there should be a ban of certain ways of treating offenders plays a crucial role in legal contexts. For instance, under the Eighth Amendment to the US Constitution offenders have a right to be free from cruel and unusual punishment. And Article 3 of the European Convention on Human Rights prohibits 'in absolute terms torture or inhumane or degrading treatment or punishment'. These prohibitions have been the subject of much legal dispute (see e.g. Ryberg 2004). However, the purpose in the following is not to consider whether a neurointervention imposed on an offender as a punishment lives up to existing legal requirements. The purpose is only to consider whether there are reasons to hold that such an intervention should be morally banned as a type of punishment.

Second, even though many penal theorists subscribe to the view that there are certain types of treatment and punishment that should never be inflicted on offenders because they are cruel or unusual, violate dignity, are inhumane, or degrading, the question as to whether a particular type of punishment falls into any of these categories is not straightforward. In the penal theoretical literature, there are various ways of treating offenders which have been held to always be unacceptable with reference to these categories. The examples range from horrendous treatments such as boiling an offender, eviscerating him, or various kinds of corporal punishment, to other severe but less horrifying treatments such as being subjected to arbitrary strip searches, having to defecate in a confined space in front of strangers, having one's head shaved, being forced to forego the expression of political observation, or being arbitrarily deprived of access to reading or to religious materials. However, what is striking is that in many cases the claim that such treatment violates an offender's dignity or is degrading or inhumane is not sustained by more precise conceptual explications of what these terms means. In some cases, it even seems that the claim that a particular treatment constitutes a violation of dignity (or some of its relatives) functions almost as a conversation stopper rather than as a fully elaborated argument.[15] Obviously, this is theoretically unsatisfactory. It may imply that such claims become either postulatory or, even worse, that they end up as rationalizations in cases where a treatment strikes one as unacceptable (rather than working as arguments from which the wrongness of the treatment is derived). Thus, ideally, a discussion of whether a certain treatment violates dignity, is inhumane or degrading should begin from a conceptual clarification of these terms followed by arguments establishing that a certain treatment falls under the extension of the concepts. In the following, I cannot here engage in a full discussion along these lines. Rather, in accordance with the previous discussions, the purpose will be to examine whether there are reasons to believe that punishment *qua* neurointerventions should be rejected on the ground of some of the explications that have been given of the suggested epithets.

Moreover, a final conceptual comment should be made. Some theorists have rejected the use of punishments that are degrading or inhumane, others have directed attention to punitive treatment that violates dignity or is cruel and unusual, and in some cases the concepts have been used to define each other; for instance, punishment is inhumane if it violates dignity (this has also been the case in the legal discussions).[16] However, the important thing

[15] Helga Kuhse has made the same observation with regard to references to dignity in modern discussions of bioethics. She even claims that the vagueness of the term and the pathos it inspires seem to be part of its attractiveness and that the term often hinders rather than contributes to a rational resolution of ethical problems (Kuhse 2000, p. 63).

[16] For instance, the European Court of Human Rights has in *Pretty v the United Kingdom* defined 'degrading' treatment as treatment that shows 'a lack of respect for dignity' (2002, p. 35 EHRR 1 at para 52).

here is the basic idea of punishment that should be banned in principle. Thus, for reasons of ease in exposition I will, in the following, talk about a banned-in-principle punishment or, in short, a BIP-punishment. The question then is whether a punishment *qua* neurointervention constitutes a BIP-punishment.[17]

Treating humans as a means only. A first possible account of a BIP-punishment that should be briefly mentioned draws on Kant's well-known distinction between means and ends. Kant introduces the distinction in relation to this second formulation of the categorial imperative—sometimes referred to as the 'dignity principle' (see e.g. Hill 1992)—according to which, man is to be treated as an end in himself and never as a means only. As such, man possesses 'unconditional and uncomparable worth', which is contrasted with what has a 'price' or only conditional value. Anything with a mere price has a value that 'admits of equivalents', which means that it may be a subject of calculated trade-offs (Kant 2012, section 2). Kant draws on this distinction in his rejection of the utilitarian approach to punishment, which treats the offender merely as an instrument or, as he puts it, as 'a means subservient to the purpose of another' (Kant 1887, p. 195f) Following this idea, could it be held that a neurointervention that is forced upon an offender constitutes a BIP-punishment because it treats the offender as a means only?

The question as to what is precisely meant by treating someone as a means only is far from simple (see e.g. Hill 1992, chap. 10; Parfit 2011, chap. 9). However, in the present context we need not engage in a thorough discussion of this issue. The reason is that the idea of treating someone as a means only is concerned with the *purpose* for which a person is treated in a certain way, that is whether the person is treated instrumentally or non-instrumentally to some end. However, what we are considering here is whether there are some *types* of treatment that should never be inflicted on offenders, which is a different question. In other words, if, as argued above, it is correct that a forced neurointervention can function as a punishment, and if it is imposed on an offender because he or she has committed a crime,

[17] There are some interpretations of BIP-punishments which will not be considered in the following. For instance, it has sometimes been suggested that what it means for a punishment to be cruel and unusual (partly) is that it is disproportionately severe (see e.g. Murphy 1979, p. 235). However, as we have seen above, there is no reason to hold that a punishment *qua* a neurointervention need violate the principle of proportionality. Another aspect that will not be considered any further is the view that a punishment should be opposed in principle because it is 'unusual'. I agree with Shaw that it 'is not clear why the state should be restricted to 'traditional' forms of punishment as opposed to 'unusual' ones' (Shaw forthcoming, p. 17).

then this would not amount to a purely instrumental treatment of the offender (see also Ryberg 2018). On the contrary, the offender gets what he deserves.[18] It might perhaps be objected that, even if the neurointervention is imposed as a punishment, the mere fact that this particular type of punishment—due to its desirable consequences—is chosen rather than other types, is sufficient to classify and reject it as instrumental. However, this is not plausible. What matters on the Kantian account, and more generally from a retributivist point of view, is that the punishment is imposed because it is deserved, that is, because the offender has committed a crime. As long as this is the case, there is no reason to abandon a punishment just because it also has desirable consequences. (Recall that the Kantian view does not rule out the possibility of treating another person instrumentally, as long as he or she is not treated as a means *only*). Thus, the idea of rejecting neurointerventions as constituting BIP-punishments, on the grounds of a moral prohibition against treating someone as a means only can easily be dismissed.

Causing a feeling of inferiority. A second explication of the view, that there are types of punishment that should be opposed in principle, has been to direct attention to the way in which some punishments are experienced by the subject who is punished. For instance, this view has been defended by Graham Zellick who defines degrading punishment as 'punishment that is inescapably humiliating and debasing beyond the normal limits of punishment, such that it reduces the essential humanity and dignity of the victim, leaving him with a feeling not simply that he has suffered discomfort or inconvenience or worse as a result of wrongdoing but that he has been reduced in status and subjected to an indignity incomparable with the status of man' (Zellick 1978, p. 669). What is appealing about this view is that it refers to the way some sorts of punitive treatment will be *experienced* by the subject undergoing the treatment.[19] *Prima facie* it seems reasonable to hold that there are some types of punishment that should never be imposed on an offender because they are in some way experienced as much more terrible than standard types of punishment. Thus, the question is whether an attempt to characterise a BIP-punishment along these lines is sufficient to rule out

[18] Given the assumption, of course, that the punishment is not disproportionate to the gravity of the crime.

[19] Such an account of degrading punishment has also been presented in legal contexts. For instance, in *Ireland v UK* it was held that '[d]egrading treatment or punishment . . . focuses on the notion of humiliation; being treatment which arouses in the victim "feelings of fear, anguish and inferiority"' (see Lawson 2005, p. 464).

forced neurointerventions as an acceptable type of punishment. In my view, there are two reasons to be sceptical of this view.

The first reason has to do with the more general question, as to whether the fact that a punishment causes the types of feeling to which Zellick refers constitutes a satisfactory justification of the conclusion that such punishment should never be imposed on offenders. There is of course no doubt that it is terrible for an offender to experience being debased or to be made to feel inferior. However, it is also a fact that standard types of punishment, such as imprisonment, cause highly unpleasant emotional reactions. Thus, in order for this account of a BIP-punishment to work, one would have to explain why the sort of feelings to which Zellick directs attention implies that such a punishment should never be imposed on an offender, while standard punishments such as imprisonment that cause various other types of highly unpleasant feelings are regarded as fully acceptable (in fact morally required). For instance, why is it unacceptable to cause a feeling of inferiority in an offender for a short period, while it is acceptable to cause a strong feeling of loneliness or usefulness for a long period of time, say during a five-year period of incarceration? To my knowledge, such questions have not been properly addressed by adherents of an experience-based account of BIP-punishments.[20]

The second and more important reason concerns the question whether it is actually the case that a forced neurointervention causes the type of experiences in offenders that should never be caused by an acceptable type of punishment. As noted, there is no doubt that such compulsory measures will usually be experienced as highly unpleasant. But is it also the case that a neurointervention will result in a sufficiently strong feeling of indignity or of being inferior? For instance, will the fact that one is forced to go through an intervention, partly because one suffers from severe problems of aggression or impulse control, cause the feeling that one is reduced in status (in a way that is significantly different from the feeling caused by incarceration)? Or, will the imposition of an intervention designed to increase certain cognitive capacities or lower the influence of particular highly uncontrollable preferences result in the experience that one is being subjected to an 'indignity incomparable with the status of man'? I must admit that I do not know if there is

[20] It might perhaps be suggested that the reason why particular types of punishment should never be imposed is that they simply cause a much stronger or more powerful—and hence worse—experience than standard types of punishment. However, this answer emphasizes the significance of the other problem considered below.

any evidence supporting these claims. However, my guess is that there are no simple answers. Rather, these will depend upon the precise nature of the intervention, the duration of the period in which the offender is going through the treatment, and probably the extent to which it is explained to the offender why he is forced to go through this particular type of punishment (and, obviously, there is no reason that offenders should not be properly informed). But if this is correct, then the suggested account of a BIP-punishment will hardly provide a *general* ground for rejecting the use of neurointerventions as a type of punishment.

Taken together, what the previous considerations amount to can be presented in a standard dilemmatic form. Either, one will have to hold that an experience-based account of BIP-punishments forbids punishments that result in feelings that are morally much more terrible or severe than feelings caused by standard punishments, such as longer terms of incarceration. However, in this case, it becomes less obvious that such feelings will generally be caused by punishments *qua* neurointerventions. Or, one may subscribe to the view that a BIP-punishment not only rules out punishments which cause terrible experiences but also feelings that are highly unpleasant albeit to a less extreme extent. In this case, it becomes more likely that several types of punishment *qua* neurointerventions should be categorized as BIP-punishments. However, the same is likely to be the case with more severe versions of standard punishments, such as imprisonment. Therefore, in short, given the way the experience-based account of BIP-punishments has been presented it is not clear that it manages to establish that compulsory neurointerventions belong to the type of punishment that should be banned in principle—at least, not if one accepts standard types of punishment.

Overwhelming capacities for rationality. While the approach that we have just considered focuses on the emotional reactions or feelings that a punishment may arouse in offenders, a very different view—though one maintaining the focus on the impact on the offender—would be to hold that a BIP-punishment is one that involves an unacceptable influence on the offender's cognitive capacities. We have already seen, above, that some penal theorists underline that a proper punishment must address an offender as a rational being. However, even if one does not subscribe to the view that a punishment in order to be justified must fulfil certain rationality requirements that go beyond the minimal necessary conditions presented in the definition of punishment, one might still hold that a punishment is unacceptable in principle if it in a deleterious way interferes with or undermines the offender's rationality.

Several theorists have emphasized the significance of human rationality as crucial in the explication and justification of the existence of BIP-punishments. For instance, John Kleinig holds that cruelty involves the infliction of suffering that threatens to overwhelm our capacity to reflect on options and to make choices based on considerations of those options (Kleinig 1998, p. 282). In the same vein, Jeffrie Murphy strongly objects to the use of various types of torture by contending that a person who is subjected to such treatment cannot 'be expected to understand this while it goes on, have a view about it, enter into discourse about it, or conduct any other characteristically human activities during the process' (Murphy 1979, p. 233). But what precisely does it mean to hold that a punishment constitutes a BIP-punishment if it overwhelms the offender's capacities for rationality?

From an overall point of view, the contention can be interpreted in different ways. The first possibility would be to hold that a punishment should be banned in principle if it overwhelms an offender's possibility of *exercising* his rationality capacity. This could very well be the case if unbearable pain is inflicted on an offender. For instance, Murphy contends that 'sending painful voltage through a man's testicles to which electrodes have been attached or boiling him in oil, or eviscerating him, or gouging out his eyes—these are not human ways of relating to another person' and they all place the subject of the treatment in a position in which he is deprived of the possibility of having a reflective attitude to what is happening to him (Murphy 1979, p. 233). Another, and more radical interpretation of the view, would be to hold that what matters is not that a treatment prevents the possibility of exercising rational capacities but rather that it destroys or diminishes such capacities; for instance, some types of torture that have been proved to have a diminishing long-term effect on a subject's cognitive capacities (e.g. the capacities to reflect on or recollect matters). And it is possible to imagine a treatment that may permanently diminish or even destroy some capacities for rationality (e.g. some types of medical treatment or, of course, the death penalty). But given these two interpretations, are there reasons to believe that a neurointervention imposed on an offender will work in either way?

It is hard to see why this should be the case. First, there are no reasons at all to believe that a neurointerventions would destroy or diminish an offender's capacity for rationality.[21] As has been noted several times, we can imagine some types of intervention which are designed to improve rational capacities. And even when interventions do not work in this way, there is no reason to believe that they will have any deleterious effects on such capacities. Neither are there any reasons to believe that neurointerventions would constitute a problem in relation to the first interpretation, that is, that they will influence the offender's possibility of exercising rationale capacities. For instance, contrary to the infliction of pain, a treatment that involves the injection of drugs or a non-invasive stimulation of the brain does not imply that the offender's ability to exercise rational capacities is affected. As far as I can see, the only situation in which it makes sense to maintain that an individual's rational capacities are overwhelmed would be a more extreme case where the person is forced to go through some sort of neurosurgery and, therefore, is placed under general anaesthesia while the intervention is performed. In this case, it is obviously correct to hold that the possibility of exercising rational capacities (or perhaps the capacity itself) has been ruled out. However, if an adherent to this approach to BIP-punishments is having this sort of temporary overwhelming or ruling out of rational capacities in mind, then it would be of value to know why even such a short obstruction of rationality is sufficient grounds for banning a punishment. After all, if a person is able to engage in all types of rational considerations before and after the neurointervention has been performed, then is it also crucial that the person is able to do so at the very moment in which the intervention is carried out? In my view, the suggested approach to BIP-punishments presupposes a plausible answer to this question (which, to my knowledge, no adherents of the view have given (see also Ryberg 2018). However, I will not here pursue this question any further. As we have already seen, this interpretation of the BIP-punishment seems to be relevant only when the imposition of an intervention implies that an individual is temporarily placed in a state of unconsciousness which, after all, is usually not the case.[22]

[21] Though, of course, it should be kept in mind that there is a risk that a certain treatment can go wrong and that this thereby—as an unintended effect—could result in deleterious consequences on some of the treated person's cognitive capacities.

[22] It might be suggested that BIP-punishments comprise all punishments that cause alterations in the mind of the offender, and which thereby also affect the offender's approach to the crime he has committed and his view on the punishment that is imposed upon him. However, this very broad interpretation of a BIP-punishment seems to coincide with the previously discussed view of the right to a freedom of mental self-determination, which means that it will be vulnerable to the same objections.

Representing offenders as sub-humans. A third explication of the idea that there are some types of punishment that should never be imposed on offenders has sometimes been referred to as the 'representational model' (e.g. Duff 2005; Ryberg 2018). While the two accounts we have just considered share the focus on the impact that a punishment may have on the offender—either emotionally or in terms of rational capacities—the representation model holds that it is unacceptable to impose a punishment that *portrays* the offender as a being of lower status. A punishment must never treat a person *as if* he or she were sub-human.

This approach has sometimes been held to have the advantage that it is not contingent on the way in which a punishment influences the offender. If what matters morally is how a punishment is experienced by an offender or the extent to which it overwhelms an offender's rational capacities, then it follows that there may not be a *general* background for banning certain types of punishment. Given the fact that there are interpersonal differences with regard to how sensitive each individual is to different types of punitive treatment, it could be the case that a punishment would constitute a BIP-punishment relative to some individuals and yet not to others. However, by focusing on the portrayal that a punishment depicts of the offender, the representational model avoids this sort of implication. Furthermore, it has been suggested that the representational model in a more plausible way manages to deal with the fact that individuals who have over a long period of time been subjected to degrading treatment may get so used to it that they fail to perceive or experience the degradation. The representational model, allegedly, can explain why such treatment is unacceptable despite the absence of the experience. The point here, however, is not to examine the relative merits of different accounts of BIP-punishments, but to consider whether the suggested model provides a sufficient background for rejecting forced neurointerventions as an acceptable type of punishment.

An indication of the fact that this account might have implications with regard to neurointervention is found in Shaw's discussion of the topic. Even though her approach does not aim at providing an explication of BIP-punishments, she nevertheless objects to the use of some types of forced neurointervention by holding that 'subjecting a person to direct brain interventions would amount to *treating her as if* she was a puppet, an automaton or a robot' (Shaw 2014, p. 12). Following this view, is it plausible to contend more generally that the imposition of a punishment *qua* a neurointervention would constitute a BIP-punishment because this would portray the offender as a puppet, an animal, or sub-human?

This question cannot be straightforwardly answered. The problem is that it is far from obvious what precisely it means to portray someone as sub-human. Thus, in order to consider whether the imposition of a neurointervention would fall into this category some conceptual clarification is required. Moreover, as noted, an explication of such a view would also be necessary in order to examine whether this account of a BIP-punishment is consistent with the fact that standard types of punishment are regarded as morally acceptable. For instance, one would have to explain why placing someone behind bars does not constitute a BIP-punishment—even though this sort of treatment is one that humans have often reserved for animals. Thus, how can this view be interpreted?

An initial answer could be that a punishment constitutes a BIP-punishment if—contrary to standard types of punishment—it fails to address the offender in a communicative manner. As we have already seen, some theorists have underlined that a punitive reaction that consisted solely in the imposition of suffering would treat the offender as nothing but more than an animal. Whereas a reaction which conveys an appropriate condemnatory message to the offender would address him or her as a responsible human being. However, this condition has already been discussed above. As we have seen, it is hard to defend the view that a properly imposed neurointervention cannot serve a communicative function. What this indicates, then, is the fact that what matters with regard to the interpretation of the representation model cannot be that a forced neurointervention constitutes a BIP-punishment due to its inability with regard to communication; rather the argument must be that neurointerventions treat offenders as sub-humans precisely because of the contents of the message that is sent to the offender (and perhaps to other parties).

The second and more plausible interpretation of the model, therefore, is that a punishment treats an offender as if he were sub-human insofar as it actually sends this sort of message to the offender. But does the imposition of a neurointervention send such a message?

At the end of the day, this is an empirical question and, to my knowledge, no one has provided any evidence to this effect. Also, it is hard to see why we should believe it to be the case. Consider the following situation. An offender, who has committed crime X, is sentenced to one year in prison during which he has to undergo a forced neurointervention by taking a neuropharmaceutical that is designed to increase his impulse control. Taken together, this period of incarceration and the intervention constitute the punishment for X.[23] That this punishment will be unpleasant for the offender is obvious. But this is not what we are considering. The question here is: Does this punishment convey the message to the offender (or the society at large) that he is an animal or sub-human?

It seems far from clear that this is the case. Could it not be the case that the punishment conveys a message along the lines: 'You have done wrong by doing X and the reason we impose this particular type of punishment on you is that we do not want you to fall back into crime in the future'? Needless to say, the point of this question is not to suggest that there do not exist types of punishment that will portray offenders as sub-humans. In the gruesome repertoire of punishments that have been (and still are) imposed on offenders, there are various methods that have been designed specifically to convey a particular picture of the offender (e.g. punishments that specially serve the purpose of presenting offenders in humiliating situations or outfit). However, obviously this is not what is at stake with regard to the imposition of neurointerventions. Thus, the point is simply that there are reasons to be cautious when it comes to the claim that neurointerventions convey such a message, if this is not backed up empirically. And that *prima facie* it is far from clear why we should believe it to be the case.

But is this contention too hasty? Is it not at least the case that the fact that a neurointervention is designed to alter a certain aspect of the offender's mental life sufficient to generate the message that the offender is not functioning in a fully desirable manner? That is, that he or she is not fully human? In a superficial sense this is true. The reason that a neurointervention may be selected as a proper punishment over other types of punishment is that it is expected to cause or facilitate some sort of reform in the offender. However, the question is whether such a message is sufficient to sustain the view that this punishment constitutes a BIP-punishment. What adherents have often objected to are punishments which are clearly degrading or humiliating. But once again it is not clear that the message along the lines that 'you suffer from a problem of impulse control, therefore, for the future benefit of others (and perhaps yourself), we impose this treatment on you' belongs in this category.

[23] To object that the combination of imprisonment and a neurointervention cannot amount to a single overall punishment is not plausible (see note 9).

Moreover, if all one is asserting is that the infliction of a neurointervention will send a message that an offender is altered in a certain way, then it is not obvious that this message will carry the same moral weight as the message conveyed by a punishment which is deeply degrading or humiliating, and which adherents of BIP-punishment have usually strongly opposed. In other words, if one lowers the bar for the kind of messages it is unacceptable to send to offenders (or others), then it becomes equally less obvious that one will be able to plausibly sustain the view that such punishments should be banned in principle. For instance, to contend that it can never be justified to serve more healthy food in prison if this is done partly because it has been shown to prevent problems in prisons or to facilitate participation in rehabilitation programmes, and that the reason is that this would send the message that the offender is improvable or not perfect, is hardly convincing. And it will certainly become more difficult to buttress the view that such a treatment is always unacceptable, while at the same time maintaining that it is fully acceptable to place offenders behind bars for years. Thus, it is not clear that the latter move can be made without running the risk of draining the representation model of some of its moral force. Therefore, given the fact that the former explications of the model left it, at best, as an open question as to what would follow with regard to neurointerventions, it seems fair to conclude that much work still needs to be done if one wishes to show that such interventions would constitute an unacceptable type of punishment on the grounds of the representational model.

Thus, what kind of conclusion is warranted on the grounds of the considerations in this section? What we have seen is that the well-known thinking that there are some types of punishment that should never be imposed on an offender has been given different interpretations. The question that arose was whether a punishment *qua* a neurointervention would fall into this category. However, it is important to underline what I do *not* believe has been shown in the discussion of this question. The point has not been to suggest that neurointerventions cannot be imposed in ways that are degrading, inhumane, cruel, dignity violating, or in other ways that have usually be used to substantiate the claim that certain punishments should be banned in principle. More modestly, the conclusion is that it is far from clear that the imposition of neurointerventions on an offender need fall into any of these categories. Thus, the contention, as presented by several theorists, that a neurointervention will constitute a BIP-punishment seems premature.

Neurointerventions and the Significance of Penal Substitution

Suppose that the considerations we have discussed so far in the chapter are all correct. That is, suppose that it is the case that a neurointervention, if properly imposed by the criminal justice system, can function as a type of punishment and, furthermore, that such interventions need not fall into the category of punishments that should be banned in principle. What does this mean with regard to sentencing? What is the moral significance of this conclusion?

The answer is that this depends upon what kind of penal theory one believes justifies the imposition of punishment on offenders. If one subscribes to a consequentialist theory or some sort of mixed theory punishment, then the question naturally is whether neurointerventions would in some cases be morally preferable to other types of reaction to criminal behaviour. But suppose instead that one does not believe that much room should be left for considerations of consequences in penal theory. More precisely, suppose that one believes that a punishment is justified as a deserved reaction to crime and that offenders should be given no more and no less than the punishment they deserve (i.e. the one that is proportionate to the gravity of the crime). Seen in this light, does the fact that a neurointervention could function as a punishment not constitute a rather modest result? Could it not plausibly be held that the criminal justice system already possesses a sufficient range of penal options to deal both with those who commit the most serious crimes and those who commit crimes at the other end of the scale, and therefore what all the previous considerations accomplish is to suggest yet another type of punishment to a system which already possesses what is required to ensure that justice is observed? In short, in this view, is the previous discussion not most properly characterized as a comprehensive conceptual exercise with very little or no genuine moral significance?

The answer to this question is in the negative. The conclusion—that a proper reaction to the suggestion that a neurointervention can function as a punishment is not merely an indifferent shrug of one's shoulders—follows from a fact that can easily be established, namely, that respect to future consequences should play a role in all plausible theories of punishment. That respect to consequences plays a crucial role in several penal theories is no surprise. Even if one does not subscribe to a pure consequentialist theory, there are many theories that imply that consequences should play some role within certain retributivist constraints. But what if one subscribes to a full-blown retributivist theory? What if one holds, as Michael Moore, that the 'desert of an offender is a sufficient reason to punish him' (Moore 1997, p. 88). Does it then follow, as Moore seems to believe, that a reference to 'consequences is simply beside the point' (1987, p. 111)? That this is not the case can easily be shown. Even if desert provides a sufficient justification for punishing offenders and the principle of proportionality is sufficient to determine the gravity of a punishment, this does not answer the question as to *how* the offender should be punished. Consider a situation in which two different types of punishment would be equally severe, that is, they would both constitute a proportionate punitive reaction to a particular crime. If that is the case, that is, if the penal bite can be preserved if one substitutes the one type of punishment for the other, then it seems that there is room for considering which punishment will have the best consequences (see also Ryberg 2019). In fact, not only is there such a room within the retributivist theory: the contention that one should in such cases not take future consequences into account would commit one to the view that *everything else being equal* it does not matter if one can prevent citizens from becoming victims of crime. But this seems morally highly implausible. In short, if it is possible to prevent future crimes without in any way compromising justice, then this is something to which I suggest even the out-and-out retributivist would be morally committed.[24]

The only way in which this conclusion can be blocked is if the retributive theory in itself dictates what constitutes the proper *type* of punishment. But it is hard to see why this should be the case. As far as I can see, the only possibility seems to be to subscribe to *lex talionis*, according to which what was done to a victim should in a strict sense be reversed against the offender. However, even though Kant comes close to defending this principle when he insists that a proper principle of penal distribution can assign 'both the quality and quantity of a just penalty', there are to my knowledge no modern

[24] Though limited attention has been directed to the discussion of types of punishment in the modern retributivist tradition, some retributivists have actually defended the idea of penal substitution (see von Hirsch et al 1989). The most recent discussion of penal substitution, to my knowledge, has been presented by Douglas Husak who contends: 'Suppose two kinds of punishment each impose Q [the deserved quantum] and thus treat offenders as they deserve. If the first is much more expensive to taxpayers than the second, it would be ludicrous to hold that anyone (retributivist or otherwise) must be indifferent between them' (Husak 2019b).

penal theorists who subscribe to a literal interpretation of 'like for like'.[25] And, indeed, *lex talionis* has been shown to face serious problems.[26]

But if a retributive theory of punishment does not dictate the type of punishment that should be imposed, then this may well have implications with regard to the use of compulsory neurointerventions on offenders. If it is the case, as has been argued, that a forcible neurointervention can function as a type of punishment, and if neurointerventions in terms of crime prevention (and other possible future consequences) are preferable to other types of punishment, then it seems to follow that when such interventions can preserve the penal bite prescribed by the proportionality principle, it would, even from a wholehearted retributivist outlook, be morally desirable to substitute traditional types of punishment with neurointerventions. It is clear, of course, that this conclusion hinges on a number of empirical assumptions. However, the point here has simply been to underline that the possibility of using neurointerventions as a type of punishment may not only have implications for theories that have been designed to take consequences into account but also for full-blown retributivist theories.

Conclusion

What we have seen in this chapter is not only that there are reasons for considering whether neurointervention can possibly function as a type of punishment, but also that the reply to this question lies in the affirmative. More precisely, it has been argued that 1) given what is often presented as the standard definition, there is no reason to believe that a properly imposed neurointervention cannot satisfy what constitute the necessary and jointly sufficient conditions for a sanction to function as a punishment. Furthermore, several theorists have held that, in order for punishment to be morally justified, it must satisfy a number of conditions that go beyond those

[25] Though Kant is often presented as an adherent of *lex talionis*, it is also the case that he suggests several types of punishment which violate the definiteness of this principle literally understood (see Kleinig 1973, p. 121f; Ryberg 2004, p. 125).

[26] An obvious objection that has been directed against *lex talionis* is that it suffers from the defect that it does not allow for the mental state of the offender to play a role in the determination a proper punishment (see Ten 1987, pp. 151–152; Kleinig 1973, pp. 120–123). Another more practical objection has been that it is often impossible to reverse a crime against the offender. For instance, as Kleinig points out, it is hard to see what punishment should be imposed on 'a rapist, a blackmailer, a forger, a dope peddler, a multiple murderer, a smuggler, or a toothless fiend who has knocked somebody's else tooth out' (Kleinig 1973, p. 120). See also the discussion in Ryberg (2004, chap. 4).

that are held to define something as a punishment. However, as we have seen, 2) it is reasonable to believe that the extra requirements that have been suggested can be satisfied by a properly imposed neurointervention. Moreover, in considering which punishments are morally justified, it is a widely accepted view that there are various types of punishment that should never be imposed on offenders. Some theorists explicitly hold that neurointerventions would fall into this category. However, as argued, 3) given the different accounts of what characterizes punishments that are degrading, cruel and unusual, inhumane, or dignity violating, there is no ground for a general moral ban of the use of punishments *qua* neurointerventions. Finally, it has been argued that 4) the fact that at least some neurointerventions can function as punishments is a conclusion with normative implications for all penal theories. Since considerations of the consequences of a punishment should play a role in all theories (though of course to a varying extent), the conclusion would have implications even for the wholehearted retributivists; namely, in relation to the question of what constitutes the morally desirable type of punishment.

But if all these considerations are correct, then what do they show? Has the ground been cleared for drawing the conclusion that, given the assumption that there currently exist some neurointerventions—or, more likely, that there will in the future be developed some—that will be preferable to other types of punishment in terms of crime prevention, one should start to use such interventions in the punishment of offenders? The answer to this question, I believe, is in the negative. What the previous discussion shows is that some neurointerventions could *in principle* be used as a punishment. However, the move from such theoretical considerations to a conclusion about the application of methods of punishment in actual penal practice opens up a plethora of further complicated issues. As an anticipation of what will be argued later, one might adhere to the view that some neurointerventions should in theory be used as a punitive reaction to criminal offending, while at the same time holding that there are strong reasons against the implementation of such punishments in current penal practice. What reasons could possibly buttress this position is something that we will return to in Chapter 7. However, before we reach this point in the investigation, there is first, as the following chapter will show, another important question that arises in relation to considerations of the use of neurointerventions in penal practice.

5

Neurointerventions and Physician Participation

Neurointerventions can take various forms. As we have seen, such interventions may vary with regard to the degree of invasiveness, their duration, the effects they are designed to produce, and the side-effects they might cause. However, it is clear that (almost) all the different types of brain and mind modifications mentioned in the previous chapters, and which are being discussed in the theoretical debate, share a common feature. The administration of these interventions presupposes the involvement of medical expertise.

In this respect, the use of neurointervention as an instrument for crime prevention does not differ from the use of such interventions in standard clinical healthcare; though, of course, the purpose is different. In a criminal justice context, the administration of the interventions requires the identification of those individual offenders whose future behaviour is likely to be affected by a treatment; assessments of possible side-effects; determination of the precise nature of the treatment (e.g. what would constitute proper dosages or duration of a treatment); and the effectuation and continued surveillance of the treatment. All this requires competences that reach far beyond those usually possessed by professionals in standard criminal justice practice, who, roughly speaking, are limited to the administration of a proper punitive treatment of offenders. However, this prompts a traditional and much debated question, namely, what role should healthcare professionals play in criminal justice practice? For instance, and more narrowly, is it acceptable for physicians to be involved in the administration of neurointerventions to criminal offenders? Or, could it possibly be the case that the use of neurointerventions would be morally blocked by, on the one hand, the fact that such treatment presupposes medical expertise and, on the other, that it would be morally wrong for physicians to offer this expertise?

The purpose of the present chapter is to consider this highly controversial question. Unlike many of the issues previously discussed, the question concerning physician participation in neurointerventions within a

Neurointerventions, Crime, and Punishment. Jesper Ryberg, Oxford University Press (2020). © Oxford University Press.
DOI: 10.1093/oso/9780190846428.001.0001

criminal justice context has, as yet, received almost no theoretical attention. However, given the fact that the use of such interventions as a criminal justice instrument is conditional on medical involvement, it is plain that the question should be addressed here. Moreover, even though discussion of the proper role of physicians in criminal justice practice may seem a minor digression relative to the cardinal question as to whether it can be justified to influence the brains of offenders, the two questions are clearly interrelated. As indicated, there may be little point in discussing whether and under which circumstances neurointerventions can be justified should it turn out that it would simply be wrong for those possessing the skills to engage in this treatment to administer it. Moreover, as we will see, some reasons against the involvement of physicians can in themselves be reasons against the use of such interventions. Furthermore, as will be argued, the question concerning the proper role of physicians may be dependent on the answer to the question whether the infliction of neurointerventions on offenders is morally legitimate. Thus, the two questions are in several ways theoretically connected.

The point of departure of the discussion in the ensuing sections will be a related issue that has received comprehensive attention, namely, physician participation in the death penalty. The purpose will not be to take a stand on this controversial question. Rather the point is to examine whether any of the objections that have been advanced against physician participation in this type of punitive practice may also constitute reasons against such participation in the use of neurointerventions on offenders. More precisely, we will consider arguments concerning the moral significance of the Hippocratic oath, the moral relation between physician and patient, the implications of the 'heal' and 'do no harm' prescriptions, and the importance of the general trustworthiness of physicians. As will become clear, I do not believe there to be a simple and clear-cut answer as to whether it is morally acceptable for physicians to take part in a crime preventive use of neurointerventions. Some of the reasons against this kind of involvement hinge on empirical questions on which it is fair to say that we do not possess sufficient knowledge. However, this also means that the reasons cannot just be dismissed. Furthermore, it will be argued that, contrary to what several commentators have held in relation to the death penalty—namely, that the unacceptability of physician participation is independent of the morality of the death penalty—I do not believe that the moral status of such participation can be separated from whether the use of a certain criminal justice procedure is morally desirable. This means that there will be some cases in which it is wrong for physicians to provide the kind of expertise which the use of neurointerventions presupposes, but also that there is no general reason to believe that the involvement of physicians will morally block the use of such interventions in the criminal justice system. To see this, it will be helpful to get a brief picture of the controversy surrounding participation in the death penalty. This is where we will start.

Physician Participation and the Death Penalty

It is a well-known fact that physicians professionally take part in various types of non-clinical-care activities. These activities include involvement in, for instance, medical research, forensic psychiatry, and military and occupational medicine (Litton 2013). Characterizing some of these types of involvement is the fact that they reach beyond, and in some cases apparently conflict with, the traditional medical aims of restoring and preserving the health of the individual. Striking and controversial examples are physician-assisted development of harsh physical and psychological interrogation techniques or participation in torture (e.g. Clark 2006). In a criminal justice context, non-clinical activities have also taken various forms. However, there is little doubt that the sort of involvement that has, at least in the more recent era, prompted most reactions and discussion is physician participation in capital punishment.

Throughout modern history, physicians have played an important role in the development of techniques to perform capital punishment. A famous example is the decapitation technique that was advocated by Dr Joseph Guillotine during the French Revolution, and which subsequently came to bear his name. Later, the electric chair was developed and introduced into the United States by Dr Carlos MacDonald and Dr E. C. Spitzka, both of whom attended the first electrocution (Ragon 1995). Execution by asphyxiation was also developed with medical expertise. However, it is the use of the lethal injection that most directly made physicians integral to the killing process. After the termination of the execution moratorium in the United States in the mid-70s, the idea of execution by intravenous drugs was suggested by a chief medical examiner and a professor of anaesthesiology in Oklahoma (Groner 2008). The first execution by lethal injection was carried out in Texas in 1982. With the climbing execution rates that followed in the two decades after the first execution, the lethal injection became standard across the United States.[1]

The introduction of new methods of execution have often been supported by the argument that the new technique was more 'humane' than previous methods. For instance, in contrast to torture-executions (such as 'breaking on the wheel' and 'burning at the stake'), Joseph Guillotine endorsed the decapitation technique as a 'quick and painless' way of killing (Ragon 1995, p. 976). The electric chair was also conceived of as a method that would cause an instant and painless death (Groner 2008, pp. 887–888). Likewise, the lethal injection was introduced as an 'extremely humane' execution method as compared to previous methods (see Groner 2008, p. 890). However, as this technique became more widespread it also initiated a heated debate. In contrast to other methods, the three-drug sequences of the lethal injection cocktail that became standard procedure implied what has been called a 'medicalization' of the death penalty (see e.g. Clark 2006; Walker 2017). Medical skills are critical to the successful performance of this procedure, for instance, with regard to correct vein location, the setting up of infusion sets, the correct selection and administration of appropriate drugs, the monitoring of the procedure and finally, the declaration of death (Groner 2002; Walker 2017). In several ways, these methods mimic standard procedures performed in clinical healthcare. Moreover, several states have, as part of their capital punishment statutes, introduced requirements concerning the presence of physicians at executions (Ragon 1995). Together, these facts have contributed to the rise of the debate as to whether physicians should be involved at all in this sort of procedure.

A glance at the recommendations presented by physician and other medical professional organizations reveals that views on this question have been remarkably univocal. After the first lethal injection was scheduled, the World Medical Association (WMA) presented a clear resolution against physician participation. In a later, revised, version it was held that: 'It is unethical for physicians to participate in capital punishment, in any way, or during any step of the execution process, including its planning and the instruction and/or training of persons to perform executions' (WMA 2008). Along the same lines, the American Medical Association (AMA) has stated that: 'A physician, as a member of a profession dedicated to a preserving life when there is

[1] For a brief description of some of the more recent changes and problems with regard to drug supplies, see Nelson and Ashby (2011, pp. 30–31).

hope of doing so, should not be a participant in a legally authorized execution' (see Walker 2017, p. 4). Corresponding statements have been made by the World Psychiatric Association, and by several other organizations representing different groups of healthcare professionals (Groner 2008).

Given the fact that physician participation in executions has unanimously been proclaimed unacceptable, it has of course also been urgent to specify more precisely what is regarded as involvement. At this point, as indicated in the above statement from the WMA resolution, the general view has been that participation comprises almost any kind of involvement. For instance, as part of the explication of 'participation', AMA has held that: 'In the case where the method of execution is lethal injection, the following actions by the physician would also constitute physician participation in execution: selecting injection sites; starting intravenous lines as a port for a lethal injection device; prescribing, preparing, administering, or supervising injection drugs or their doses or types; inspecting, testing, or maintaining lethal injection devices; and consulting with or supervising lethal injection personnel' (see Groner 2008 p. 904). In fact, the only acceptable involvement would consist in a pre-execution prescription of sedatives when requested by the offender, and certification of death after it has been pronounced (Rothman 1995; Walker 2017).

The purpose in the following is not to examine the question whether the professional codes of medical ethics that address capital punishment also have implications with regard to a criminal justice use of neurointerventions; rather, as noted, it is to investigate whether any of the arguments that commentators have presented to back up the opposition to physician participation in execution can plausibly be held to apply with regard to neurointerventions. As indicated by the more precise explications of what physician participation implies, there are clear parallels between the use of lethal injections and at least some types of neurointervention. As will become clear, however, there are also dissimilarities.

The Legalist Argument

The first argument that should be mentioned holds that physician participation in the death penalty is morally wrong because a lethal injection, and probably also other kinds of involvement in executions, would be contrary to the professional code of ethics of physicians. Though several of the statements and resolutions that have been presented by healthcare organizations are not strictly binding on physicians, they nevertheless provide a code of conduct for members of the profession.[2] And, as we have seen, they clearly oppose physician participation in the death penalty. However, an even more common background for rejecting involvement by physicians has been to refer to the Hippocratic oath. As we will see shortly, the answer to the question as to whether physician participation in the death penalty conflicts with the Hippocratic oath is not so simple. However, rather than engaging in a discussion of the contents of the oath or of certain codes of conduct, one could start by asking whether the fact that a certain type of physician involvement is prohibited would in itself constitute an argument against the involvement. Turning to our discussion of the use of neurointerventions, the argument would be that physician participation in a criminal justice use of such interventions is morally wrong because this participation contravenes the Hippocratic oath or perhaps other professional codes of conduct. If we *arguendo* assume the factual premise of this objection to be true, would an argument along these lines be convincing?

The problem facing this legalist argument is that a mere reference to the contents of the Hippocratic oath or a certain code of conduct does not settle the *moral* question. Rather the argument seems to constitute a *petitio principii* by presupposing the wrongness of the particular practice which it is supposed to establish. Put in a standard dilemmatic form: Either there are good moral reasons supporting the contents of the oath or a code of conduct but, in that case, these reasons constitute the grounds for rejecting physician participation in criminal justice use of neurointerventions, not the oath or the code themselves. Or, there are no moral reasons buttressing the contents of the oath or a code of conduct, in which case the fact that the use of such interventions conflict with the contents of the oath or the code of conduct does not constitute a ground for rejecting involvement in the use of neurointerventions. Either way, it seems that the legalist argument cannot in itself constitute a convincing objection against this sort of physician involvement.

However, it might be objected that analysing the legalist objection in this way is too simple. More precisely, it might be suggested that physicians are

[2] For instance, though the American Medical Association can revoke a physician's membership, the organization has no legal enforcement authority (see Groner 2008, p. 904). However, it should be mentioned that the American Board of Anaesthesiology has ruled that anaesthesiologists will not be certified if they participate in capital punishment (see Nelson and Ashby 2011, p. 29).

morally bound by the prohibitions. This contention seems most relevant in relation to the Hippocratic oath.[3] After all, an oath is not just a set of guidelines to which an individual can decide to adjust her behaviour, rather is it something to which the individual who has taken the oath has committed herself. Gerald Dworkin, as part of his considerations and rejection of physician involvement in lethal injections, has defended a view along these lines. What he suggests it that 'to the extent that a code is incorporated into an explicit oath taken by the professional, this, like any other promise or commitment, must at least be relevant to her obligations' (Dworkin 2002, p. 184). Could one thus hold that physician participation in criminal justice use of neurointerventions is morally wrong, in so far as this conflicts with the content of the Hippocratic oath by which physicians are bound? There are, I believe, several reasons why this reference to oath-taking cannot settle the issue.

The first question raised by this version of the legalist argument is, of course, to what extent is one morally bound by having taken an oath? As is well-known, the views on this traditional issue differ significantly. Some would hold that there are instrumental reasons for keeping an oath. Generally, it is important for social life and practice that we maintain trust in each other; when it comes to specific promises, there may be more precise instrumental reasons in favour of acting in accordance with what has been promised (at least, under normal circumstances). Others defend a traditional non-instrumentalist approach to the way one is morally bound by an oath. With such a view it would be morally acceptable to breach the oath only when sufficiently strong countervailing reasons are at stake. Or, one might subscribe to an absolutist view which does not allow for breaches no matter what. I shall not enter here into the traditional moral dispute between the different approaches to oath-taking, except to remark that it is not clear that the first two approaches are inconsistent with physician participation in criminal justice use of neurointerventions, and that the final approach opens up a number of highly counterintuitive implications. In the present context, there seem to be a number of more practical reasons that strike me as sufficient to establish that the legalist objection cannot settle the question about physician participation.

[3] However, perhaps it may also be suggested that a physician who has not taken steps to object to a professional code of conduct is morally bound by this code by having tacitly accepted it (see e.g. Dworkin 2002). Whether the moral assumption behind this view is plausible is not something I will pursue here. But it should be noted that, contrary to the question of the death penalty, it is not clear that physician participation in neurointerventions would be prohibited by existing codes of conduct.

The problem is that, perhaps contrary to popular belief, physicians' commitment to the content of the Hippocratic oath is no simple matter. First, a study conducted by Robert M. Veatch and Cheryl G. Macpherson on oath-taking at a US medical school indicates that medical students who take an oath may not always share the values expressed by the oath. According to the two researchers the 'problem of medical school oath-taking is that schools that pick one of the existing codes . . . will be inflicting a particular ethical view on a large percentage of students who do not start out with such commitments' (Veatch and Macpherson 2010, p. 344). But this raises the question as to how strongly one is morally committed to an oath one is asked to recite upon entry into a profession, if one does not even fully believe in the content of the oath.[4] However, we need not pursue this question here. Second, as the quote by Veatch and Macpherson indicates, there is no shared oath taken by all physicians. The practice of oath-taking varies from one country to another, and even within one and the same country. For instance, in the United States some medical schools use the Hippocratic oath, others use the AMA code, while some draft their own oaths. And the differences between these oaths may be substantial (see Veatch and Macpherson 2010, p. 343). But this means that some physicians *qua* their oath-taking may be bound to a certain practice to which others are not bound. Third, and most importantly, there are many physicians who have not even taken an oath. For instance, Veatch and Macpherson mention there have been periods in England when oath-taking was uncommon (Veatch and Macpherson 2010, p. 341). What this shows is that, even if one subscribes to a Kantian rigorist interpretation of the extent to which one is morally bound by taking an oath, this will not suffice as a general argument against physician participation in criminal justice use of neurointerventions.

In sum, even though references to the Hippocratic oath have been very frequent in the discussion of physician participation in the death penalty and could therefore also potentially be expected to be of relevance to the discussion of the use of neurointerventions as regards offenders, a legalist interpretation of such references—according to which the Hippocratic oath in itself answers the moral question concerning physician involvement—is

[4] To suggest that one is not bound by an oath if one does not really believe in the contents of the oath is hardly plausible. However, there could perhaps be degrees with regard to the extent to which one is bound by an oath depending on the degree to which one honestly and sincerely believes in what one is promising. As noted, I will not pursue this question here. As we will see, there is a more important objection to the legalist argument.

not persuasive (see also Litton 2013; Walker 2017). The content of the oath varies and some physicians may not even have been required to take an oath. Finally, it should be recalled that the discussion of the legalist objection was based on the assumption that the content of the oath, or a code of conduct to which (some) physicians are committed, would actually prohibit participation in treatment of offenders by neurointerventions. As we will now see, it is far from obvious that this is the case.

The Heal and No Harm Objections

From having considered the legalist question as to whether physicians are morally bound by the Hippocratic oath or by a code of medical ethics, we can now turn to the content of the oath itself and to such codes. Central to the discussion of physician-assisted capital punishment have been the views, first, that such involvement contravenes the basic obligation of physicians to promote health and, second, to not cause harm. Both views are clearly stated in the Hippocratic oath. According to the oath, physicians commit themselves to 'use treatment to help the sick'. And the oath begins with the dictum *primum non nocere*, that is, 'first, do no harm' (see e.g. Ragon 1995). Likewise, AMA has strongly underlined both obligations in its rejection of physician participation in capital punishment. As Lee Black and Robert M. Sade observe: 'The central thread running through the AMA's Code of Medical Ethics is the physician's obligations to help and not to harm people. The result of an execution, however, clearly harms the executed person without offsetting benefit—no rationale can justify a different conclusion—so physician participation in executions is manifestly unethical' (Black and Sade 2007, p. 2780). Whether it is actually the case that these two obligations imply non-participation by a physician in the death penalty is not the question here. However, since both goals seem to constitute plausible obligations for physicians, the question is whether these obligations succeed in establishing the moral wrongness of physician involvement in criminal justice use of neurointerventions. Let us consider each obligation in turn.

That physicians have an obligation to heal and promote the health of patients seems uncontroversial. Moreover, though neurointerventions can of course be used for purely clinical purposes, this is not what we have been considering in this book. If a neurointervention is used for crime preventive reasons, say, as a way of improving impulse control in an offender suffering from an explosive temperament, or as a method for reducing the influence of particular paraphilic preferences, then it seems that we have moved far beyond what would usually count as medical healing or health promotion. However, it would be an obvious fallacy to infer from the obligation to heal patients the wrongness of participating in treatment for crime preventive purposes. The inference would have no more validity than the claim that because a lifeguard has an obligation to save people from drowning, he or she has an obligation not to help someone who has been run down by a car. Obviously, the latter claim does not follow from the former. Therefore, in order for the objection to be valid, one would have to subscribe to a much stronger premise, namely, that physicians should not apply their medical skills for other purposes than those narrowly associated with healing and the promotion of health. In fact, this seems to be the view advanced by some commentators in the death penalty debate. Several theorists have held that, by using medical skills for non-medical purposes, the physician no longer acts as a physician or that the physician would thereby, in an unacceptable way, move beyond the purposes and goals that ultimately define medicine.[5]

However, while the view that physicians have an obligation to heal patients, as mentioned, seems highly plausible, the stronger premise that it is wrong to use medical knowledge and skills for other purposes than healing, is much harder to sustain. Once again, it is possible to imagine purely instrumental reasons that could perhaps substantiate this premise; and, as noted, we will return to such a reason shortly. However, to suggest that a physician for non-instrumental reasons would be acting wrongly by using medical skills for purposes that go beyond the healing of patients is not a tenable position.

First, it is a fact that physicians currently use their medical skills in various ways that seem morally acceptable and which do not constitute healing or health promotion (e.g. autopsies). In fact, sometimes these activities are more appropriately characterized as enhancements (e.g. plastic surgery). And, physicians may even be part of the work of the criminal justice system (e.g. when giving expert testimony in court). Thus, the view that the work of physicians should be limited to healing, and that participation in the criminal justice use of neurointervention reaches far beyond acceptable physician

[5] For instance, as part of his criticism of physician involvement in the death penalty, Pellegrino claims that, 'Even if the patient saw the doctor's participation as beneficent, the act of killing controverts the healing purposes of medicine and would have to be refused' (Pellegrino 1993, p. 376). For a critical discussion of this view, see Litton (2013) and Walker (2017).

practice, seems highly questionable in the light of the activities in which physicians are already engaged. Second, and more importantly, the main problem with the healing-only view is to explain why the work of physicians should be thus narrowly constrained. If there are cases in which the application of medical knowledge and skills—even if it cannot be subsumed under the standard activities of healing and health promotion—may have desirable consequences, then why should physicians not engage in such activities? It requires a strong argument to explain why the use of medical expertise should not be expanded to such cases. And the problem is, that it is hard to see what such an argument would amount to. Merely insisting that such activities do not count as healing or health promotion is not satisfactory if non-participation happens at the cost of something that is morally desirable. Therefore, the absence of an argument buttressing the view that physicians should be concerned only with healing, and the fact that physicians often engage in activities that reach beyond this narrow scope of obligation—and do so in a way that is hard to challenge morally—clearly indicate that the healing argument is unsatisfactory.

But if the healing-based objection to physician involvement in crime preventive use of neurointerventions is not persuasive, what about an objection based on the do-no-harm principle? At first sight an objection premised on this medical *dictum* seems much more promising. That physicians should not cause harm to others is certainly morally appealing. Moreover, as we have seen, the imposition of neurointerventions on offenders may in various ways be harmful. In fact, for retributively minded theorists who would insist on the view that the proper reaction to crime is punishment, the justified use of forcible neurointerventions would, as we have seen, be contingent on the fact that such interventions are in fact harmful. Therefore, the premises that physicians should not cause harm and that neurointerventions would usually be harmful seem to provide a simple ground for a conclusion stressing the non-participation of physicians in the use of such interventions. However, on closer scrutiny things are more complicated.

To suggest that the do-no-harm principle implies that it is morally wrong, simpliciter, for physicians to cause pain or other types of suffering to patients, is clearly absurd. It is a well-known and unavoidable effect of many kinds of clinical treatment that they involve suffering. Medicine has various types of side-effects. And some kinds of medical procedure, say surgery, may be painful even if all appropriate steps are taken to alleviate this effect. An obvious reason to hold that such effects do not conflict with the do-no-harm *dictum* is that a patient (hopefully) will benefit from the treatment. The suffering that is caused by the treatment will be less than the suffering that would in the future have befallen the patient had he or she not been cured of the disease. However, if this sort of balancing is consistent with the do-no-harm principle, then this will also affect the objection which the principle allegedly constitutes in relation to the use of neurointerventions. If a neurointervention is harmful—but, everything considered, benefits the offender—then this treatment will not be blocked by the principle. And, as we have seen, it is not unthinkable that a neurointervention could have such an effect. A person may benefit significantly, and in various ways, from not falling back into crime. Thus, in this interpretation it seems that the principle does not prohibit physician involvement in neurointerventions.[6] In fact, even if we disregard the possible future beneficial effects of such interventions, there is an even more simple way in which the treatment could be held not to harm an offender. If we have a case in which an offender either gets a traditional punishment, say imprisonment, or a neurointervention (or perhaps a reduced time of incarceration combined with a neurointervention), then insofar as the neurointervention is less harmful than the standard punishment, the offender could be held to have benefited. To contend that this sort of benefit violates the do-no-harm principle, even though it is preferable for the offender compared to the alternatives, would render the principle highly dubious. Various types of clinical treatment place a patient in circumstances involving less suffering than those the patient would have faced without the treatment.

Some might at this point object that the above considerations rest on a simple misunderstanding of what the do-no-harm principle implies. What the principle is concerned with is not whether or not someone is harmed or benefited, everything considered. Rather, the principle's central distinction is whether harm is inflicted intentionally or knowingly. More precisely, what the principle prohibits is that physicians intentionally inflict harm on others. This is the reason why painful clinical treatment is consistent with the principle; the suffering that is caused is only a foreseen side-effect. However, such an interpretation of the do-no-harm *dictum* gives rise to several questions.

[6] It is worth noting that, if one subscribes to the view that a punishment is defined (partly) by being locally harmful—that is, all that matters is that, say, a month in prison is harmful while it goes on, not whether everything considered it makes an offender's life worse (see Chap. 4)—then it is possible that a neurointervention could function as a punishment (by satisfying this harm-requirement). This it could do without violating the do-no-harm *dictum* (if the intervention happens to benefit the offender, everything considered).

First, it is worth mentioning, that things sometimes become more complicated when one starts talking about intentions in institutional contexts. For instance, one might wonder whether the principle would still be consistent with a case in which the criminal justice system sentences an offender to a neurointervention which is intended to function as a punishment, and hence be harmful, but where the physician who is administering the intervention does not intend the harm. However, there is no need to further pursue this question.

Second, and more importantly, if the do-no-harm principle is interpreted as a principle against intentional harm, there are two ways of understanding this. First, the principle should be regarded as a standard deontological constraint against harm-doing. But, in that case, it seems morally highly arbitrary why only physicians should be bound by such a constraint if it exists. It seems reasonable to expect that it would be equally binding for all other parties, including those who usually execute a punishment, and for the state itself. But this means that the principle would ultimately end up blocking the use of state punishment, which is not an implication many are willing to accept.[7] Alternatively, it could be suggested that the principle should not be interpreted as a constraint, but rather as one instrumentally designed to prevent some future morally undesirable consequences. In this interpretation, the do-no-harm principle will not challenge the existence of a punishment institution. There could be instrumental reasons as to why physicians should not engage in some sorts of harm-doing even though it is fully acceptable (or even morally required) for the state and those parties usually authorized by the state to engage in harmful punitive treatment of offenders. However, in this interpretation, the objection collapses into a purely instrumental justification; that is, it becomes conditional on whether there exist such instrumental reasons in favour of excluding physicians from engaging in harm-doing. As indicated, we will return to this question shortly.

Third, and most importantly, it should be noted that, even if it were the case that it could be explained in a non-arbitrary manner why only

[7] It might of course be held that there is a constraint against *undeserved* harm. But, in that case, the constraint would not forbid the use of a neurointervention if this, as considered in the previous chapter, is administered as a deserved punishment. Another possibility is to hold that there is a threshold such that, when there are sufficiently strong countervailing reasons, the constraint will be overruled; consequently, the existence of such a constraint does not imply that punishment is unacceptable. However, if one subscribes to a threshold position, there is no reason to ban physician participation in neurointerventions on the grounds of the do-no-harm *dictum*, as long as the countervailing reasons are sufficiently strong.

physicians as a special group should be bound by a constraint against harm-doing, this would still not be sufficient to exclude physician participation in criminal justice use of neurointerventions. While this constraint might exclude the use of such interventions as a punishment which—as we have seen previously—according to standard definitions involves intentional harm, there is no problem if such interventions are used by the criminal justice system but in a non-punitive way. That is, neither if an intervention is used as a compulsory (but non-punitive) instrument to crime prevention nor if it, less radically, is offered to an offender as an alternative to punishment, would it be the case that the principle would be violated. In neither case would the harm that accompanies the treatment be *intended*. Therefore, the objection based on the do-no-harm *dictum* cannot be regarded as persuasive.

In summary, it seems fair to conclude that, irrespective of what the implications of the heal-only prescription and the do-no-harm principle would be with regard to physician participation in the death penalty—which is not our topic—it is not the case that the two principles provide a sufficient background for morally excluding the involvement of physicians in criminal justice use of neurointerventions.

The Patient-Oriented Objection

Another argument, frequently presented in discussion of physician participation in capital punishment, is based on the view that such involvement constitutes a failure to properly acknowledge *whose* interests the physician should ultimately be protecting. By engaging in punishment, the argument goes, the physician will no longer be acting out of concern for the individual patient—whose interests ought to have primacy—but rather on the grounds of the interests of the state. For instance, Edmund Pellegrino has argued that actions such as not only the injection of lethal drugs, but even the functioning as a witness to an execution, or pronouncing death upon order, would all 'make the physician the agent of direct and intentional killing, a non-beneficent act, which is in the state's, and not the patient's, interest' (Pellegrino 1993, p. 376). In the same vein, Lee Black and Hilary Fairbrother—both members of AMA's Council on Ethical and Judicial Affairs—have held that: 'This is the very essence of the patient-physician relationship: The physician acts for the benefit of the patient. We cannot ignore the patient-physician relationship between a prisoner and physician participating in an execution by simply characterizing it as something other than what it is, a physician using his or her medical expertise on an individual' (Black and Fairbrother 2008, p. 60). Their view is clearly represented in the AMA principles of medical ethics, which explicitly underline the physician's responsibility to the patient as paramount. Several other commentators have raised similar objections. Thus, the question here is whether a reference to the special responsibility of physicians to patients suffices to buttress an objection against physician involvement in crime preventive use of neurointerventions on offenders. In simple form, the argument would be that such involvement would be wrong because it constitutes a way of pursuing state interests rather than patient interests.

Though the argument has some affinity with the healing-based objection, which we have just considered, it could also be held to have a comparative advantage. As we have seen, one of the problems of the healing-based objection is that it depicts an implausibly narrow picture of what constitutes the purpose of a physician's work. However, this perspective is widened by focusing more broadly on the patient-physician relationship.[8] Moreover, it seems appealing to underline the significance of this relationship. However, be that as it may, it seems to me that such an argument against physician participation in neurointerventions is vulnerable to some of the same types of objection.

The first thing to be noted is that, on closer inspection, this objection can be given two very different interpretations (which have not always been distinguished in the death- penalty debate). First, it could be held that it is wrong for physicians to act out of other interests than those of their patients. If this premise is combined with the assumption that offenders are *not* patients, and that crime preventive use of neurointerventions is simply a way of pursuing societal interests, we are led to the conclusion that physician participation in neurointervention on offenders would be wrong. However, the problem with this version of the objection is that the view that it is wrong for a physician to act out of other interests than those of patients is hard to accept. As a simple example, a physician may decide not to give antibiotics to a patient out of consideration of the general danger of antibiotic-resistant bacteria (see Litton 2013). And this could be the case, even if the decision implies that it will take longer for the individual patient to recover from a disease. Moreover, as already noted, it is generally accepted that physicians take part

[8] For instance, the patient-physician relationship may comprise enhancement of patients, or other types of physician-patient interaction, which are not usually regarded as healing.

in research which may sometimes not be of benefit to the individual patient. In fact, in some cases research may even expose a patient to (minor) risks.

It might perhaps be objected that these examples are unconvincing for the simple reason that, even though they involve decisions by physicians that are not in the interest of the individual patient, they are nonetheless performed with the interests of other patients in mind. The attempt to avoid antibiotic resistance is a way of protecting future vulnerable patients. And, more generally, engagement in research may in the future promote the interests of patients. However, even if this is correct, there are various other ways of acting that cannot be explained thus. As pointed out by Litton, a simple example is that '[o]n a given day, a physician could be quite able to continue providing high-quality care to her patients yet have strong family-related reasons to leave the hospital' (Litton 2013, p. 341). And even if this is done only when the physician has a colleague to cover her patients, the shift itself may increase the risk of medical errors and may therefore not be in the best interests of the patients. However, such shifts seem acceptable. Litton's more general conclusion is that 'despite lofty talk of physicians' duty to individual patients and health in general, we plainly accept non-health-related reasons as relevant to the scope of physician obligations' (Litton 2013, p. 341). I believe that Litton is right on this point; more generally, the problem with the first interpretation of the objection is simply, that it is morally very hard to justify the view that it is wrong for physicians to do something that would benefit the public or the state even if this could be done at no cost to others such as individual patients.

A second and very different interpretation of the patient-based objection would be to hold that it is wrong for physicians to pursue state interests at the cost of patient interests, and to combine this premise with the assumption that offenders should be regarded as patients. While the first interpretation is based on the view that involvement in the treatment of offenders is not a patient-oriented activity but only a way of pursuing state interests, the second perceives offenders as a group of patients whose interests are sacrificed in the attempt to realize broader societal interests. As the initial quotes indicate, it is apparently this interpretation of the patient-based objection that critiques usually have had in mind in their opposition to physician participation in the death penalty. However, this interpretation also faces serious problems.

As was the case with the first interpretation, the second also conflicts with what is usually regarded as the reasonable obligations of physicians. For instance, physicians may in particular cases have a moral (and legal) duty to sacrifice the confidentiality of a patient in order to protect the interests of third parties or the public in general.[9] Or physicians may decide to quarantine a patient out of societal interests, even if this is against the interest of the individual patient. However, leaving such examples aside, the main flaw of the objection is that, even if we accept that societal interests should not be pursued at the cost of patient interests, this does not provide a sufficient ground for the conclusion that physician involvement in neurointerventions on offenders is morally unacceptable.

The problem is that the use of neurointerventions in the criminal justice system need neither be harmful nor conflict with the interests of those who are being treated. This is precisely what we have already seen in the above rejection of the do-no-harm objection. If we have a case in which an offender can choose between a standard punishment and a neurointervention then, given these alternatives, it could well be in the offender's interest to choose the latter. In particular, this is likely to be the case if the neurointervention is much less unpleasant than the punishment, such as would be the case if the state wishes to offer an incentive to get the offender to accept the treatment (see Chap. 2). Thus, the pursuit of societal interests need not be contrary to patient interests.

It might perhaps be objected that this cannot count as a plausible way of discussing neurointerventions as being in the interest of the offender. After all, it is most likely to be the case that the offender would prefer to have neither the neurointervention nor the punishment. Therefore, in this sense, the physician would still be acting against the interests of the offender. However, this is not a plausible response. If I accept that a physician performs an operation on me in order to cure me of a disease, then it would be absurd to contend that this would be against my interests, or that this would harm me, simply because I would prefer to have neither the operation nor the disease. Thus, what is in the interest of an individual must be determined on the ground of the actual alternatives available to the individual.

Some might perhaps feel that this comparison is inappropriate for the simple reason that, when it comes to the choice of the offender, it is the state that has set the scenery by making the punishment the only available alternative to the offender, which means that the state could have provided other alternatives. In the case of the operation, the state has obviously not had a

[9] For an instructive discussion of the implications of physicians' obligation to report certain infectious diseases even when this disclosure violates patient confidentiality, see Litton (2013).

role to play. The suffering from the disease would be the only genuine alternative to the operation.[10] In other words, one might fear that when the available alternatives are determined by the state, then this would legitimize any kind of physician involvement (e.g. even in torture), as long as this treatment is less terrible than the one the criminal justice system would have otherwise imposed on the offender. In my view, this objection is not convincing. If the treatment offered by physicians would be preferable to the individual, and if this treatment constitutes the only possible alternative to a terrible treatment imposed by the state, then I tend to believe that physician participation would *ceteris paribus* be morally acceptable.[11] Moreover, it should be noted that if one rejects this view, then one will have to give an account of what the relevant kind of non-comparative harm, which physicians should never cause, consists in; and one will have to do so in a way that is consistent with the fact that physicians may sometimes non-comparatively cause severe harm to patients as part of clinical treatment. However, what is more important is that, in the present context, one need not even accept the suggested implication. In order to ensure that physician participation in neurointerventions is unaffected by the patient-based objection, all one has to accept is that there are at least some cases in which the use of neurointerventions does not harm the offender. And it is hard to believe that this is not the case.

Where then do these considerations lead us? What we have seen is that the patient-based objection against crime preventive use of neurointerventions can be interpreted somewhat differently. It could be held that the scope of the obligations of physicians is limited to patients in the sense that it is wrong to engage in any non-patient-oriented exercise of medical expertise. However, as we have seen, this view runs counter to what is already an accepted part of the work of physicians, and it is morally very difficult to explain why it would be wrong for a physician to use his or her skills in something morally desirable just because this does not involve patients. Alternatively, it could be held that physicians should never pursue state interests at the cost of the interests of their patients. However, there are cases indicating that even this

[10] For instance, Brody and Wardlaw make this point in relation to the discussion of participation in the death penalty: 'The prisoner is constrained within a total institution; the inevitability of his death is imposed on him by societal, human judgment, and not by the natural workings of any disease' (2008, p. 53).

[11] In practice, the *ceteris paribus* clause would probably not be satisfied. For instance, it could be the case that there are instrumental reasons—such as the trust-based reasons to which we will return below—why physicians should not engage in terrible ways of treating offenders, even when the treatment constitutes the only and preferable alternative to the state treatment.

view could be questionable; more importantly, as argued, this interpretation of the premise would not be sufficient to block at least some ways of using neurointerventions on offenders for crime preventive purposes. In the absence of other possible interpretations, it seems reasonable to conclude that the patient-based objection is not convincing.

The Trust-based Objection

While the previous arguments focus on something that is regarded as inherently wrong in the application of medical expertise in non-clinical-healthcare practice or, at least, in a criminal justice context, a different type of argument has been presented, namely, that physicians should stick to their last, that is, focus on healing, on their relationship with patients and, at least, observe prescriptions against participation in the death penalty, for instrumental reasons. Though this argument can take different forms, there is no doubt that the most frequent concern advanced in the debate is that, once physicians transgress the usual domain for the exercise of medical expertise by being involved in the executions of offenders, this will erode trust in physicians. For instance, Brody and Wardlaw reject physician participation by contending that: 'The professional integrity of the physician must be grounded in the basic fact of the patient's deep vulnerability, and the overriding importance of maintaining patient and public trust' (Brody and Wardlaw 2008, p. 54). Along the same lines, Ragon strongly opposes physician participation in lethal injections by underlining that: 'Physician and patient trust is a key aspect of the medical relationship, and any physician action that harms the ability of a patient to trust their doctor could lead to a breakdown in the medical profession' (Ragon 1995, p. 999). The fact that many commentators have strongly emphasized the significance of trust in physicians as a major concern in relation to capital punishment makes it obvious to ask whether the same concern is appropriate in relation to physician participation in crime preventive use of neurointerventions. Briefly, the objection would be that the involvement of physicians in criminal justice use of neurointerventions is wrong because this will undermine trust in physicians. Is such an argument persuasive?

In order to examine this objection, one will have to ask whose trust is at stake, why we should expect there to be an effect on the trust placed in physicians, and why this trust should be regarded as important. In relation to the death penalty, answers to these questions have not been univocal. Some commentators have underlined the significance of how physician participation will undermine the trust of prison inmates. For instance, Groner notes that '[m]any patients would not feel comfortable knowing that his or her doctor might be an executioner' (Groner 2008, p. 906). And Ragon argues that, when physicians aid in the execution process, inmates 'do not see the physicians and other health care workers as healers who are concerned about their well-being. Rather, the inmates perceive the physicians and other health care workers as people who will someday kill them. Inmates who are not on death row complain about this conflict since they must submit to care of the same physicians. The physician may have participated in the execution of one of their peers' (Ragon 1995, p. 1000). A possible implication of a breakdown in an inmate's trust in physicians could be that he or she fails to reveal important symptoms to the physician and, consequently, does not receive the required treatment (see Groner 2008, p. 905). That this would be highly undesirable is indisputable (in particular in the light of the sad fact that a high number of inmates are suffering from physical and psychological impairments (see e.g. Enggist et al. 2014; Senior and Shaw 2007).

However, it has also been suggested that the erosion of trust might well reach further. Several theorists contend that involvement in the death penalty erodes public trust in the medical profession as a whole (see Brody and Wardlaw 2008; Groner 2008; Lerman 2006). For instance, it has been underlined that physicians who assist in executions often do so anonymously and they may, at the same time, practice in hospitals. This implies that patients in general could fear being treated by someone who is also an executioner (see Groner 2008, p. 906). Furthermore, the way in which the use of lethal injections mimics standard clinical treatment may in itself constitute a cause of mistrust. The president of the American Society of Anaesthesiologists has underlined this: 'The more the execution looks like an anesthetic, the less comfortable patients are likely to be with anaesthesia. Surgery is already a frightening time and one in which patients need to trust their anaesthesiologist. The last thing patients need is to equate the [operating room] with a death chamber, to equate anesthetic drugs with death drugs, or to have in their subconscious the spectre of the anesthesiologist as an executioner' (Guidry 2006; also quoted by Litton 2013, p. 346).

The question we are faced with here is to what extent can any of these consequences to prison populations, or the public at large, be expected to follow from physician participation in criminal justice use of neurointerventions.

This question is not a normative one, but purely empirical, and the most reasonable answer in my view is that this is very hard to determine. It seems reasonable to expect that consequences of physician participation will be dependent on a number of questions concerning the use of the interventions. What types of intervention are being used? How many offenders will be treated? Are the interventions only used voluntarily or also in a compulsory manner? What is the relation between the physicians who administer the interventions and those who perform standard clinical treatment of inmates or patients in general? Thus, it is hardly plausible to expect a general answer concerning the consequences of physician participation in neurointerventions. And, unfortunately, it is hard to extract any useful lesson from the debate on lethal injections.

Though many commentators, as mentioned, have directed attention to the threat of the erosion of trust, there are also some commentators who are sceptical about the objection. For instance, Litton contends that the suggested mistrust by patients of anaesthesiologists 'may rest on a dubious empirical assumption' (Litton 2013, p. 347). And Dworkin, for his part, holds that it is not 'likely that my relationship with my doctor will be coloured by the fact that I know that either she, or other doctors, are acting as executioners' (Dworkin 2002, p. 188). In fact, Kenneth Baum even goes so far as to suggest that, since non-participation by physicians in the death penalty could be comprehended as an abandonment of the prisoners' interests, 'one could credibly argue that physicians are more likely to erode public trust by *refusing* to participate in executions than by participating' (Baum 2001, p. 69). However, what is remarkable is that generally neither those commentators who have advanced the objection, nor those who are more sceptical, have backed up their views with any empirical evidence. This, of course, means that it will be hard to make any even tentatively solid predictions of what would follow in terms of the erosion of trust if neurointervention is to be used in the future in the treatment of some offenders.

Therefore, in my view, the proper conclusion in relation to the trust-based objection is the following. When many learned scholars with medical expertise warn that physician participation in the death penalty may lead to problems concerning insecurity or mistrust in the physician-patient relationship, then there is reason to take this risk seriously also in relation to physician involvement in criminal justice use of neurointerventions. Moreover, insofar as it is the case that such involvement will threaten trust in physicians, this will not only constitute a reason against physician participation but may, to the extent that the use of such interventions is conditional on the involvement of physicians, also constitute a reason against this way of using neurointerventions. However, ultimately, this is all contingent on empirical questions which we simply do not yet possess enough knowledge to answer. Thus, for those who prefer strict adherence to or rejection of objections, this conclusion may not seem an interesting contribution. However, in my view, it is the only proper one to draw.

Finally, it should be noted that, even if some evidence can be given in support of the view that physician participation in neurointerventions will threaten trust in physicians, this cannot per se constitute a knock-down argument. The consequences of diminished trust will have to be balanced against the benefit that such interventions may have in terms of crime prevention (which, ultimately, may also concern life and health). While this balancing is less relevant in relation to the question of physician participation in the death penalty, where the preventive effect is highly dubious, it may be relevant in the discussion of neurointerventions (or at least to our present discussion of such interventions, which, as underlined, is based on the assumption that such intervention will have a crime preventive effect). As indicated above, it is therefore most reasonable to interpret the trust objection as constituting a possible reason against—but in itself insufficient ground for a rejection of—physician participation in criminal justice use of neurointerventions.

The Dependency View

A striking fact about much of the discussion of physician participation in the death penalty is that the moral status of the death penalty itself often plays no role in the assessment of what constitutes the morally right behaviour of physicians. In fact, the view that the morality of physician participation should be analysed independently of the question of the morality of the death penalty, has constituted a common view both among anti- and pro-participation commentators.[12]

For instance, organizations representing healthcare professionals have often abstained from taking a clear stand on the moral status of the death penalty. In relation to the previously cited condemnation of physician

[12] As part of his analysis of physician participation, Litton observes that the 'standard approach to the ethics of physician participation deems the moral status of executions normatively irrelevant' (Litton 2013, p. 337).

participation, AMA explicitly underlines that, 'An individual's opinion on capital punishment is the personal moral decision of the individual' (see Litton 2013, p. 336; Walker 2017, p. 4). Thus, whether capital punishment is right or wrong is believed to have no bearings on the AMA recommendations. Correspondingly, many anti-participation theorists have either not commented on the moral status of the death penalty at all, or they have explicitly underlined the importance of analysing the morality of physician participation and the death penalty separately. For instance, Lawrence Nelson and Brandon Ashby contend that: 'Unfortunately, the debate about the ethics of physician participation in executions has been thrown into confusion by the failure to separate this issue from the ethical propriety of the death penalty itself' (Nelson and Ashby 2011, p. 29).[13]

The same pattern is found in the arguments presented by theorists who have defended physician participation. The main reason in favour of the involvement of physicians in the death penalty by lethal injection has been that participation constitutes a way of preventing, or at least minimizing, the suffering of the offender. If, as part of the three-drug protocol, the initial anaesthetic is ineffectively delivered—as might happen if one does not possess the requisite medical expertise—then the subsequent paralyzing and heartbeat-ceasing agents may cause tremendous suffering to the individual being executed (Litton 2013; Walker 2017). Thus, apparently, this pro-argument works independently of whether or not the death penalty is morally justified. In fact, the moral independence of the two issues has been explicitly underlined by some physicians who have themselves taken part in executions, despite the fact that they regard the death penalty as morally wrong.[14]

But is it plausible to maintain that there is independency between the moral status of the way in which the criminal justice system decides to deal with an offender and the morality of physician participation? The purpose here is obviously not to consider this question in relation to the death penalty, but in relation to physician involvement in crime preventive use of neurointerventions. An adequate answer may open up a number of complicated questions within

[13] As the quote indicates, there are also theorists who have argued that the morality of the death penalty is crucial in the assessment of physician participation. For a brief presentation of some of these views, see Nelson and Ashby (2011, p. 29).

[14] For instance, Atul Gawande mentions the case of Dr Carlo Musso who holds that a death-penalty patient 'is no different from a patient dying of cancer—except his cancer is a court order' and, further, that he has 'the cure for this cancer'—abolition of the death penalty—but 'if the people and the government won't let you provide it, and a patient then dies, are you not going to comfort him?' (Gawande 2006, p. 1228).

political philosophy concerning the legitimacy and moral significance of state decisions, which I cannot even tentatively address here. However, what I will do is outline the contours of a way of analysing the morality of physician participation, and the rightness or wrongness of the use of neurointerventions, which implies that the two issues are not narrowly dependant in the sense that the answer to the latter question *determines* the answer to the former. But I suggest that the morality of physician participation is dependent both upon the rightness or wrongness of the use of neurointerventions, and on the influence which participation or non-participation will have on how the criminal justice system ends up dealing with offenders.

The argument in favour of the first part of the dependency view is simple; namely, that if the use of neurointerventions on offenders is morally desirable then the reasons substantiating this desirability will also be reasons in favour of physician participation. And, conversely, if this use of interventions is morally undesirable, then the reasons that make this treatment undesirable will also constitute reasons against physician participation. It is worth noticing, as we have seen in the death penalty debate, that the arguments against physician participation have often attempted to establish the wrongness of participation by showing that physicians should not transgress a narrowly defined area of obligations. This has been done by contending that physicians should be concerned *only* with healing, the promotion of health, or with their patients. However, if, as argued, these attempts at narrowly specifying the limits between right and wrong participation are not plausible, then it becomes hard to see why the reasons for or against the use of neurointerventions should not also be reasons for or against the involvement of physicians. Put in less abstract terms, suppose that there are strong moral reasons in favour of using a particular type of neurointervention on offenders, namely, that this use will prevent future serious re-offending, that it will do so much more effectively and cheaply than any alternative type of treatment, and that there will not be any serious side-effects on the offender. Why, then, should these reasons not also be the reasons in favour of physician participation? If the state has good reasons to try to effectuate such interventions, then it is hard to see why the desirability of the interventions should not also place a duty on those who possess the capacity to effectuate the treatment.[15] Of course, this does not mean that there could not also be

[15] It might be the case that further conditions should be added with regard to which physicians will be morally bound by the reasons in favour of the use of neurointerventions. For instance, it could be

some arguments against participation. For instance, there could be reasons, such as a loss of inmate or public trust in physicians. But this does not establish that the reasons in favour of the use of neurointerventions do not constitute reasons in favour of physician participation; it only shows that there could be other reasons pointing in the opposite direction.

The view that the reasons determining the assessment of the use of neurointerventions would also constitute reasons in relation to the question of physician participation, is perhaps even more appealing if we consider the opposite case, namely, if the use of such interventions is morally undesirable. It seems very hard to support the view that the wrongness of the use of neurointerventions should have no bearings whatsoever on whether physicians should be engaged in the administration of these interventions.

Thus, the fact that the rightness or wrongness of the use of neurointerventions will, in the absence of plausible arguments that succeed in determining the scope of physician obligations more narrowly, provide reasons for or against physician involvement in this type of treatment, constitutes a very simple background for the dependency view. However, as indicated, there is one further aspect that should be added to the discussion and which may influence the moral assessment of physician participation; namely, whether the administration of the treatment is conditional on the involvement of physicians. More precisely, there are two possibilities. Either neurointerventions will be administered even though physicians do not participate, or the interventions will only be administered if physicians are involved. These possibilities—the significance of which has not always been fully acknowledged in the discussion of the death penalty—complicate the dependency view in the following way.

First, suppose that the use of neurointerventions on offenders is, everything considered, morally desirable. It is possible that these interventions will be used only if physicians are involved in the treatment. Or, it could be the case that the interventions will be used even if physicians do not participate in the administration of the interventions. If the former is the case, then there are reasons for physicians to participate. If the latter is the case, the picture is more complicated. If other parties could ensure that the interventions are effectuated, then it becomes morally relevant to consider whether physicians can do the job more effectively (e.g. with less risk to the offenders)

held that it should be the physicians who are authorized by the state to engage in this sort of treatment within the criminal justice system. However, in this overall presentation of the argument, I will not consider this question in detail.

or whether there are other *pro* and *cons* (e.g. whether physician participation will result in a loss of trust which will not be caused if other parties take on the task). As noted, this is precisely what has been held in relation to the death penalty: it makes a difference whether or not physicians are involved. However, while it may be possible to conduct the death penalty without the involvement of physicians, this is, at least for many types of intervention, less obvious with regard to neurointerventions.[16] The requisite knowledge and technology is simply more complicated and, therefore, requires more competences. Thus, when it comes to neurointerventions the first situation is by far the most likely.

Second, suppose instead that the use of a neurointervention on offenders is morally undesirable. It is possible, then, that the interventions could be performed only with the involvement of physicians. If this is so, there are reasons for physicians not to be involved. Alternatively, it could be the case that such interventions will nevertheless be inflicted on offenders even if physicians do not participate. In that case, the question will be whether the involvement of physicians may imply that the interventions are administered in a less risky or otherwise preferable manner. As indicated, though, this possibility is less likely.

Thus, what is the upshot of these considerations? What I have suggested is that the moral assessment of physician participation in criminal justice use of neurointerventions should not be separated from the moral reasons that make this use morally desirable or undesirable. If there are reasons for using such interventions in the way the criminal justice system deals with offenders, then these reasons will also constitute reasons in favour of physician participation, and *vice versa* if there exist reasons against the use of such interventions. However, we have also seen that the rightness or wrongness of the use of such interventions does not *determine* the answer with regard to the propriety of physician participation, since this is also dependant on whether or not neurointerventions would be applied even in the absence of the involvement of physicians. Thus, in principle there could be cases where the use of neurointerventions is morally desirable but where it would be wrong for physicians to be involved, or where such interventions are morally

[16] Several theorists have suggested that the death penalty could be carried out without the involvement of physician. Both because other methods than the lethal injection could be applied, and because the lethal injection does not require a comprehensive medical expertise. For instance, Caplan has argued that it 'is difficult to believe that the only persons capable of administering lethal injections are graduates of medical schools', and that 'properly trained personnel should carry it out', not physicians (Caplan 2007, p. 1048).

undesirable but where it would be morally right for physicians to participate. While there are some reasons to believe that there could be cases where the death penalty can be carried out without physician involvement, I have indicated that this is less likely with regard to neurointerventions, the use of which often requires a considerable degree of expertise. Thus, in practice, the suggested dependency view may imply that the morality of physician participation is comparatively more closely tied to the morality of the use of neurointerventions.

It should be underlined again, that what I have presented here is not an argument to the effect that physician participation in neurointervention is either right or wrong. I do not believe that such a general conclusion is tenable. Rather, I have offered the contours of a way of analysing physician participation which indicate that the assessment of the question is dependent on the moral desirability of such interventions and, furthermore, on what constitute the alternatives to physician participation. As such, this view is highly contingent on both moral and empirical conditions surrounding a particular use of such interventions. The only simple conclusion is that there is no general answer when it comes to the morality of physician participation.

Conclusion

The question concerning when physician participation in the non-clinical work of the criminal justice system is morally acceptable is highly controversial and has given rise to a plethora of writings and many arguments. The considerations in this section have not aspired to present an adequate answer to this question in general nor even in relation to criminal justice use of neurointervention in particular. More narrowly, the purpose has been to examine whether there are general reasons to believe that a physician's role in the administration of neurointerventions is in itself something that would morally block the use of such interventions for criminal justice purposes; in short, whether physician participation should here be regarded as morally wrong. The comprehensive condemnation of physician-assisted lethal injections could fuel the suspicion that the use of neurointerventions would also be vulnerable to serious objections. However, as suggested, the picture is not so simple.

What has been argued is that 1) attempts at rejecting the physician participation in criminal justice use of neurointerventions by showing that medical knowledge and skills should not be used for purposes that go beyond standard clinical treatment are not plausible. Neither can the view that physicians should only take part in healing or that the use of their competences should be limited only to patients stand closer scrutiny. Furthermore, it has been suggested that 2) physician participation should be analysed in terms of what I have called the 'dependency view', which combines considerations of the moral status of the use of neurointerventions and of the extent to which the application of such interventions is conditional on the participation of physicians. Admittedly, this view is controversial and discussion of it incomplete. However, if I am right with regard to initial arguments that the scope of a proper use of medical skills cannot plausibly be very narrowly limited, then I suspect that the main criticism that would be directed against the dependency view would be that it implies too strict obligations on physicians. For instance, some might object that it is not clear that a societal benefit from the use of neurointerventions implies the existence of an obligation of physicians to realize this benefit (even though it would of course be acceptable if they attempted to do so). Or, that if the use of neurointerventions is morally wrong, then it is nevertheless not so obvious that it is wrong for physicians to be involved in the use of such interventions if this treatment is prescribed *via* legitimate political decision-making.[17] But if this criticism turns out to be correct, it will probably only support the overall attempt at showing that the criminal justice use of neurointervention is not morally blocked by the fact that this type of treatment will usually, if not always, be conditional on medical expertise.

[17] See, for instance, the criticism of moral dependency by Nelson and Ashby (2011).

6
Neurointerventions and the Lesson from History

The history of punishment is not for the squeamish. Even though the methods used to deal with criminal offenders have undergone significant changes over the centuries, all periods have had horrifying ways of punitively dealing with those who have violated societal norms or broken the law. However, the atrocities to which offenders have been subjected have not been limited to the types of punishment which states have imposed on culprits. Offenders have also been exposed to various sorts of exploitation and non-punitive maltreatment. An obvious example is the use of prisoners for medical experimentation.

One of the most horrendous examples is the atrocious experiments that were conducted during the Nazi regime on concentration camp inmates. However, it is worth noting that, when Nazi doctors were accused at the Doctors' Trial at the Nuremberg Tribunal in 1947, part of the defence consisted *not* in rejecting the fact that experiments had taken place but in underlining that the experimenters had not violated common research practice. Defence lawyers asserted that 'civilized' nations including Britain, France, and the United States had in numerous cases, often without consent, conducted dangerous medical experiments on prison inmates (see e.g. Shuster 1997). Notwithstanding the dubious quality of this assertion as a defence to the murderous and torturous experiments that were performed by the Nazi regime in the name of medicine, it is a fact that many countries had a history of experimenting on prisoners. An appalling example, cited also by the lawyers at the Nuremberg Trial, is the American malaria experiment that took place in the state of Illinois, where hundreds of prisoners were inoculated with malaria in an attempt to develop an effective way of treating this disease (Arboleda-Florez 2005). Other infamous examples that were later revealed and attracted attention are the pellagra studies led by medical officer, Joseph Goldberger, and performed partly on inmates of Ranking Farm Prison in Mississippi, and the numerous studies in which animal testicular

Neurointerventions, Crime, and Punishment. Jesper Ryberg, Oxford University Press (2020). © Oxford University Press.
DOI: 10.1093/oso/9780190846428.001.0001

substance was implanted in prisoners by the notorious Dr L. L. Stanley at the San Quentin Prison (California).

However, what is particularly noteworthy is that, despite the attention that was drawn to the Nazi atrocities in relation to the Nuremberg Trials, the practice of using offenders as human guinea pigs continued after the world war. In fact, as Allen Hornblum has emphasized in his overview article in *The British Medical Journal*, the pre-war examples of experiments on prisoners 'paled in comparison to what transpired after Nuremberg' (Hornblum 1997, p. 1437). As he notes, the fact that prisoners were 'cheap and available' made them obvious raw material for medical experimentation for several decades after the war (Hornblum 1997, p. 1437). Noteworthily, it has been estimated that, by the end of the 1960s, about 90 per cent of phase I studies in new drugs in the US were conducted on prison inmates (Arboleda-Florez 2005; Wiegand 2007).[1] This way of utilizing prisoners, which in many cases exposed those involved to serious risk, persisted until the 1970s.

As many commentators on the use of prisoners for scientific research have underlined, it is highly important to know how prisoners have historically been abused. It is important in order to understand what led to the Nuremberg Code, the Declaration of Helsinki, and subsequent research regulations and, thereby, to fully grasp why it is crucial to insist that, even when physicians as therapists are guided by the Hippocratic Oath, they should not decide alone on what constitutes ethical conduct when they work as researchers (Shuster 1997). Being cognizant of the history is important in order to learn from the mistakes of the past. But if this is the case, what follows if we turn from the broader discussion of experimentation to the narrower topic of this book, namely, the use of neurointerventions? Is there a corresponding lesson to be learned from the modern history of the use of neurointerventions on offenders? And if so, what does this lesson consist in?

A comprehensive historical overview and analysis of how neurointerventions have been used in a criminal justice context throughout the last century has, to my knowledge, not yet been presented and the present chapter certainly does not aspire to make up for this lack.[2] Rather, the purpose here is to consider to what extent previous highly unsuccessful attempts

[1] In an article from 1974, Victoria C. Heldman comments on these phase I experiments by noting: 'Such testing often results in pain, discomfort, fever or loss of weight on the part of prisoner-subjects, who are usually paid a small amount for the use of their bodies, but who are not guaranteed treatment for drug effects after the experiments ends' (Heldman 1974, p. 159).

[2] However, some highly instructive articles have been written on this issue; see, for instance, Greely (2008), Pustilnik (2009), and Shen (2016).

at manipulating the brains of offenders may constitute an objection against the present or future use of crime preventive neurointerventions. In order to consider this question, the chapter will start by outlining a few historical examples of how neurointerventions have been imposed on offenders. These examples provide the backdrop against which *the lesson from history objection* is discussed. As will be seen, authors who have emphasized the significance of the history in this field have not always been very clear with regard to what kind of conclusion is warranted by directing attention to examples from the past. The conclusion of the chapter will be, on the one hand, that the lesson from history objection is not valid in the sense that previous use of neurointervention either predetermines, or even constitutes, a moral reason against, the future use of such treatment; but, on the other, that the historical use of such interventions may serve as a highly important source of knowledge one should draw on when steps are taken from purely theoretical considerations to actual decision-making in a messy reality.

Historical Cases of Neurointerventions

Millard Wright was in his mid-30s, with a long career in crime. In the summer of 1945 he was apprehended after numerous cases of housebreaking and robbery. His case is reviewed by Edward E. Mayer in a 1948 volume of *Criminal Law and Criminology*. What made the case worth reporting was not the nature of the crimes committed, but the subsequent decision that was taken to deal with Wright's propensity for crime: Wright's counsel requested his client be permitted to have a prefrontal lobotomy in an attempt 'to cure the prisoner of his criminal tendencies' (Mayer 1948, p. 576). The petition was accepted by the judge and the operation performed.

At first sight it may strike one as surprising that a judge would grant a petition for an operation that he knew to be experimental. However, on closer inspection, the decision is less hard to understand.[3] As initially noted, experimental treatment of offenders was regarded as much more acceptable in this period. Furthermore, and more importantly, there were at the time several reports of successful attempts at using lobotomy as a crime preventive treatment. For instance, the *American Weekly Magazine* featured an article

[3] Mayer comments on the motive of the judge by saying that he 'was willing to help a human being to reestablish himself in society if it were possible to do so, though he had no precedent for granting a petition for an experimental operation' (Mayer 1948, p. 577).

stating that an incorrigible criminal 'twenty-eight years old and mentally unbalanced had been converted into a rational decent woman' (Mayer 1948, p. 577). The *New York Times* reported a case in Detroit in which a woman had spent 14 years in prison and whose record included various types of crime. Shortly after surgery the psychiatrist reported her to be 'more relaxed, more composed emotionally . . . Her former threatening, aggressive, assaultive behaviour has disappeared. In contrast to her previous defiant attitude she shows genuine embarrassment about her past conduct' (*New York Times*, 7 December 1946, p. 34). And in Pittsburgh, an offender who had spent fifteen years in prison was lobotomized and reported as cured of his 'criminal tendencies' (*Time*, 14 July 1947, p. 53). These cases are also emphasized in a 1948 article in *The Yale Law Review*, which, while acknowledging the need for improvement to the technique, ends on a very positive note by contending that: 'Perfection of so relatively simple and inexpensive a rehabilitation technique as the prefrontal lobotomy promises to be a major contribution to the cure of criminals' (1948, p. 1098).

The optimism expressed in this passage clearly reflects the prevailing view in this period of lobotomy as a cheap and safe method of curing various types of psychological impairment. Having been developed by the Portuguese neurologist, Egas Monitz, in 1935, it become mainstream psychiatric treatment in many countries (see e.g. Pustilnik 2009; Sterling 1978; Swayze 1995). Many thousands of patients were subjected to this procedure. The use of, and enthusiasm for, lobotomy peaked in the 1940s, culminating in the honouring of Monitz with the Nobel Prize in 1949. During the 1950s the use of lobotomy declined. Today this period is usually emphasized as one of the most horrendous examples of how a cocktail of underdeveloped and naïve science, and a genuine lack of other methods for treating psychological impairments, may have devastating consequences for numerous patents.[4] As Jenell Johnson has put it, lobotomy is 'remembered as one the of the most spectacular failures in the history of medicine' (Johnson 2009, p. 367). As for the crime preventive use of the procedure in the case of Millard Wright, this was not a success.[5] After the operation he re-offended and ended up committing suicide in prison (Schirmann 2013, p. 110).

[4] For considerations on what contributed to the escalating use of psychosurgery in the 1940s, see, for instance, Swayze (1995, p. 511).

[5] Interestingly, as part of his discussion of the Wright case, Mayer holds that, despite the fact that lobotomies have been performed on many thousands of patients, there 'is no justification . . . from any survey of the literature, for accepting a conclusion that criminal tendencies can be eliminated

However, in an outline of examples of how interventions in the brains of offenders have been administered as a remedy to criminal tendencies, there is no need to go as far back as the heydays of lobotomy in the middle of the last century. After a quieter decade, psychosurgery experienced a revival in the late 1960s and early 1970s.[6] This period has received less attention than the lobotomy era but was, by some commentators at the time, referred to as 'the second wave' of lobotomy.[7] As Peter Sterling, a contemporary critique of many of the brain-surgical procedures, notes, part of the motivation behind the use of psychosurgery seems to have changed in this period relative to the early days of lobotomy. While, as he underlines, the main focus of interest in the 1930s and 1940s was on the release or control of institutionalized patients, the focus of the new area 'was on control of social deviance, violence, and addiction' (Sterling 1978, p. 149). Contentions confirming this view are easy to find.

For instance, in the wake of the Detroit riots in 1967, neurosurgeon, Vernon H. Mark underlined that, even though poverty and other social factors underlie urban riots, the 'obviousness of these causes may have blinded us to the more subtle role of other possible factors, including brain dysfunction in the rioters who engaged in arson sniping, and physical assault' (Mark et al. 1967, p. 895). A more radical exposition of this view was presented by one of the very busy psychosurgeons of the time, M. Hunter Brown, who held that: 'The person convicted of a violent crime should have the chance for a corrective operation . . . Each violent young criminal incarcerated from 20 years to life costs the taxpayers perhaps $100,000. For roughly $6000, society can provide medical treatment which will transform him into a responsible, well-adjusted citizen' (Sterling 1978, p. 127). These quotes represent clear examples of the renewed enthusiasm for a reinvocation of neurointerventions as part of the armamentarium of crime prevention advocated by some scientists. In several cases, this enthusiasm was followed up by action.

by any of the several types of lobotomy which are in use. In fact, there are numerous examples cited in the literature of criminal tendencies having been initiated by this operation' (Mayer 1948, p. 582). However, Mayer does not provide references for any of these examples.

[6] However, even if this period was quieter compared to the peak period of the lobotomy era, a large number of psychosurgeries were still performed. According to some estimates, around 4000 operations were performed each year on US patients in the first five years of the 1960s (see Pustilnik 2009, p. 200, citing E. S. Valenstein).

[7] In his fine article on the history of neurolaw, Francis X. Shen refers to this period as 'an often forgotten era of brain science and law' (Shen 2016, p. 680).

In the early 1970s, plans for the use of psychosurgery in the Californian prison system were discovered.[8] However, when the plans came to light, it was also revealed that the prison system had already performed three operations in 1968 (Breggin 1975). The purpose of these operations was to destroy amygdalar tissue by the use of surgically implanted electrodes. One of the three operations was alleged to have been successful: a claim, however, which was contested by others. Peter R. Breggin mentions that lawyers who visited the prisoners 'felt that they were severely damaged by the surgery' (see Breggin 1975, p. 855). Shortly after his release, the 'successful case' was rearrested for robbery (Pustilnik 2009). The fates of the non-successful cases remain obscure.

Another example of how the criminal justice system attempted to influence the mental life of prisoners through medical procedures is the use of drugs. At the California medical facility at Vacavilla, 'anectine therapy' was—as part of an experimental treatment programme—administered as a method of curbing violent and aggressive behaviour.[9] The inmates were all convicted felons, but were found to suffer from personality or psychological problems. Fifteen inmates, described as 'angry young men', were included in the treatment (Spece 1972, p. 634). Anectine was injected into the subjects while a doctor presented verbal admonishments about the wrongful acts they had committed. The drug functioned by causing a severe physical reaction consisting in 'soreness of body muscles, prolonged respiratory arrest . . . and a state of total body paralysis often likened to death or drowning' (Spece 1972, pp. 671–672; see also Serber et al. 1975). The aim was to create an association between the wrongful act and the 'death-like' state of body paralysis as a way of generating an aversion in the prisoners to their misdeeds.[10]

Such stories of how prisoners were being treated with the use of surgery and drugs did not go unnoticed.[11] More generally, the view that criminal

[8] In a 1974 article, Jonathan Brant quotes a letter from a California prison administrator, according to whom: 'Surgical and diagnostic procedures would be performed to locate centers in the brain which may have been previously damaged and which would serve as the focus for episodes of violent behaviour. If these areas were located and it was verified that they were indeed the source of aggressive behaviour, neurosurgery would be performed, directed at the previously found cranial foci' (Brant 1974, p. 188).

[9] Another equally horrendous example of drug therapy is the anectine experiments conducted at Atascadero State Hospital between 1966 and 1969, in which over 90 male inmates were given anectine as part of an 'exploratory study' (see Serber et al. 1975, p. 89).

[10] For other examples of how drugs were used in the treatment of prisoners in this period, see e.g. Opton (1974), Sitnick (1975), or Spece (1972).

[11] The case of the use of psychosurgery was presented in the *Washington Post* of 25 February 1972 (see Aarons 1972).

tendencies (as well as other types of unwanted social behaviour) could be dealt with through psychosurgery was strongly opposed by several groups of protesters. These included academics who underlined that the procedures were not proved therapy but highly experimental, who advocated the rights of prisoners, and who recommended a general moratorium on psychosurgical operations and a reduced use of psychopharmacological agents (see e.g. Brant 1974; Breggin 1975; Chorover 1974; Sitnick 1975; Spece 1972). The treatment also initiated a broader public outcry.[12] Today no one would deny that the California cases as well as the cases from the lobotomy era are clear examples of horrendous maltreatment of offenders. They constitute powerful instances of, and thereby confirm, what Amada Pustilnik rightly refers to as the 'ill-fated and sometimes tragic affair' between neuroscience and criminal law (Pustilnik 2009, p. 183).

However, the question is what do such historical examples, which are now slowly entering the modern neuroethical discussion of the use of neurointerventions, establish? If the examples should not only be highlighted for purely historical reasons, but are also believed to have some sort of probative value for the current ethical discussion of the future use of neurointerventions, then how should such a 'lesson from history objection' more precisely be interpreted?

Neurointerventions and the Pessimist Induction

When attention to the dark history of neurointerventions is drawn into the modern debate, there is a underlying assumption that historical precedent should play an important role in the assessment of the ethical legitimacy of such treatment methods. However, beyond the contentions that it is important never to forget the history—which is hard to deny—or attempts at drawing parallels between past and present expressions of unsubstantiated enthusiasm with regard to what neurointerventions may come to achieve in terms of crime prevention—parallels which in some cases are indeed striking[13]—little has been done to make explicit how precisely the history

[12] For an instructive overview, see Casey (2015).

[13] See, for instance, Shen (2016) who holds that 'we can see many parallels between the framework of Mark and Ervin and the approach taken in a 2013 book, The Anatomy of Violence by neuroscientist Adriane Raine', and that the '1960s psychosurgery historical moment resonates with contemporary social issues in ways that are unmistaken' (Shen 2016, pp. 682–683). See also Pustilnik (2009).

should inform or influence the current ethical discussion. To my knowledge, Emily McTernan is the only one to have developed a more precise account of a lesson from history objection.

The purpose of McTernan's considerations is not to offer a conclusive case against neurointerventions 'but rather to defend the relevance of the history of this approach to tackling socially undesirable behaviour, when addressing its ethical prospects' (McTernan 2018, p. 275). History can of course be of relevance in various ways; for instance, out of concern for the compensation of numerous victims of past maltreatment. However, this is not what McTernan has in mind. Rather what she defends is a principle of 'pessimist induction', which states that: 'Previously, it has been unethical to engage in neurointerventions to target socially undesirable behavior. So we should be wary of introducing such neurointerventions into our criminal justice system now: we have a defeasible reason or presumption against engaging in these neurointerventions' (McTernan 2018, p. 278). Two things should be noted about this principle.

First, the pessimist induction not only concerns previous cases of scientific mistakes. As McTernan rightly underlines, the science of medicine, and science more generally, have been mistaken before. However, what was wrong about the use of neurointerventions in the past was 'not only a scientific mistake but also a moral one'. What matters is not only the scientific shortcomings of the use of such interventions, but the fact 'that people were wronged by these treatments' (2018, p. 277) The pessimist induction is an inference from previous *wrongful* behaviour.

Second, as McTernan has presented the pessimist induction, it is clear that the view is not that the previous wrongness of acts of a certain type predetermines the answer to a present instance of this type of act. In fact, and we will return to it shortly, this would constitute a highly implausible view. Rather, the idea is that the wrongfulness of the past constitutes a 'defeasible' reason against a present instance of an act. However, despite these apparently plausible specifications, it still seems to me that on closer scrutiny the idea of a pessimist induction does not provide a satisfactory interpretation of a lesson from history objection. In fact, the pessimist induction strikes me as morally implausible. Thus—and we will return to it shortly—though I believe that a lesson from history is highly important, the significance of this lesson is not plausibly captured in McTernan's proposal.

In order to sustain this contention, let us start by spelling out the pessimist induction more precisely. The overall idea of this inductive principle is that, if we are considering the rightness or wrongness of an instance of a particular type of act, A_0, and if this type of act has previously, in a sufficient number of instances—say A_{-3}, A_{-2}, and A_{-1}—been performed in ways that were wrong, then this constitutes a reason against performing A_0; that is, a reason that should now figure in the complete set of the rightness or wrongness making properties of A-ing. It is possible, as we will see in the next section, to interpret the pessimist induction in a very different way. However, if we take the idea of an *inductive* inference seriously then I believe this explication captures the contents of the pessimist induction. However, as will now be argued, the view faces two types of problem. First, in order to be applicable in relation to the use of neurointerventions, the view requires specifications in various ways which are quite challenging. Second, even if all the necessary specifications have been provided, the inductive inference in itself seems morally questionable.

(A) In order to determine whether the tragic history of neurointerventions should have any significance in the assessment of a treatment, there are many factors that need to be specified. First, it is necessary to know what constitutes the inductive base cases; that is, the historical cases on the ground of which the inductive inference is made. In order to identify the relevant prehistory of a present act, it is necessary to specify what *type* of act the current neurointervention is. However, it is a well-known fact that a particular act can be described as an instantiation of many *types* of behaviour. For instance, suppose the intervention in question consists in a treatment scheme involving the use of a drug that is intended to increase impulse control and that an offender has consented to this treatment. The current act could then be described, for instance, as an act of crime preventive treatment, as the infliction of a neurointervention, as the administration of a drug, or perhaps the consensual administration of a particular drug. However, if the precedence of the present act is supposed to play a role with regard to the rightness or wrongness of the act, then obviously it may make a crucial difference whether it is regarded as an instantiation of one or another of these types of act.

If the treatment is typed as an instance of crime prevention, then it will have to be assessed against the background of the whole pallet of initiatives that have historically been taken to prevent crime and, perhaps, even against the history of punishment. If, alternatively, the treatment is typed as a neurointervention, then presumably the set of relevant inductive base cases will be significantly reduced. However, the drug treatment will still have to be assessed, for instance, on the ground of a prehistory of the use of neurointervention as a standard medical treatment. The set of base cases can of course be further narrowed down if instead the treatment is typed as an instance of the infliction of a crime preventive neurointervention. But in this case, it would, for instance, still have to be assessed in the light of some of the examples of psychosurgery outlined above. Yet other ways of narrowing down the relevant set of inductive base cases would be to conceive this set of cases as crime preventive use of drugs or, resulting in an even more narrow set of cases as the act of a non-coercive crime preventive administration of a particular drug on a particular type of offender.[14]

What all these examples show is that the use of the pessimist induction presupposes that we are able to specify how a particular current act should be typed. In the absence of clarification on this point, the induction will simply not be applicable. Furthermore, the point is of course that it may make a crucial difference to the inductive inference whether the treatment is typed in one way or another. For instance, if the treatment is assessed generally in the light of the history of crime prevention, then it seems that the inductive base will include both successful and unsuccessful cases which, presumably, will undermine the inductive inference. Conversely, if it is specified in a narrower way, the pessimist induction may support the wrongness of the present treatment. If, as a further possibility, it is specified in an even more narrow way, say as the act of administering this particular drug for this particular reason, then the dark history of neurointerventions may become totally irrelevant; in fact, there may be no historical precedence for the present treatment. However serious this is, the problem of typing is only one of the challenges relating to the specification of the proper inductive base cases.

(B) Suppose that it has been determined in a morally non-arbitrary manner what constitutes the proper typing of a present instance of a neurointervention. This will provide a sufficient background for a retrospective examination of the precedents of the present act. However, in order to be able to draw an inductive conclusion, it will not be sufficient merely to identify historical instances of this particular type of act, one will also have to assess the value of each of the acts. As we have seen, McTernan holds that

[14] As indicated previously, the type of act which McTernan has in mind is 'neurointerventions to target socially undesirable behavior' (p. 277). However, she also opens for the possibility of 'grouping neurointerventions for socially undesirable behavior with eugenics and the worst elements of psychiatry' and then for a broader interpretation of the relevant type of act, namely, 'treating people, via medical interventions, for the sake of the social good' (p. 280).

the wrongness of neurointerventions in the past was not only a scientific mistake but also a moral one. But this means that more precise accounts of what constitute scientific mistakes and moral wrongs are needed. For instance, we need to specify what kind of scientific errors were involved in the use of neurosurgery on prisoners. Did the mistake consist in regarding a treatment as medically solid when this was not the case? Or, if the treatment were used experimentally—and indeed several of the atrocious examples of neurointerventions outlined above were regarded as experimental—did the scientific mistakes consist in the absence of proper experimental designs? As for the wrongness of the past use of neurointerventions, McTernan suggests that people were wronged by being exposed to risks and serious side-effects, by being victims of treatment that pursued objectionable goals, and by the fact that the inventions, relative to the intended benefits, were disproportionate in their degree of invasiveness (McTernan 2018, p. 277). Thus, in order to be able to assess the historical cases, one will need precise accounts both of what the relevant scientific mistakes consists of, and of what makes these cases morally wrong. Though I regard this as less challenging than the typing problem, it is clear that it is a task that opens up many questions. For instance, if a number of historical cases of neurointervention suffered from different scientific errors or were morally wrong for different reasons, can these cases then all still serve as the basis for an inductive conclusion or does a proper inference presuppose a high degree of similarity between the base cases?

(C) A further question that arises in relation to the inductive base is when these cases constitute a sufficient set to justify an inductive inference. Is the fact that it is possible to identify a few historical wrongful instances of an act sufficient to provide a reason against a present instance of this act? Or does a sound inference presuppose a much more comprehensive set? In relation to the use of neurointerventions, this question may sound somewhat superfluous for the simple reason that the number of people who have suffered from various sorts of neurointervention during the dark ages of psychiatry is so large. However, this question is closely bound up with the answer given to the typing challenge. For instance, if a present instance of a neurointervention is assessed (partly) on the ground of the historical use of lobotomy, then it is beyond doubt that there exists whatever constitutes a sufficient number of cases to justify the inductive inference (assuming of course that such an inference can be valid). It is estimated that in the heydays of lobotomy, up to 70.000 people were subjected to this operation in the

US alone (Pustilnik 2009, p. 200).[15] However, if what we are considering is not the procedure in general, but lobotomies on offenders, then the number is much lower. To my knowledge, precise estimates of this number do not exist. But Schirmann mentions that even in this period '[p]sychosurgery for criminals was rare and highly controversial' (Schirmann 2013, p. 110). Thus, depending on how the typing challenge is answered, the question concerning a proper size of base cases may not be totally without relevance.

(D) Suppose that all the above challenges concerning the specification of the set of base cases have been properly answered. If an inductive inference is sound, then there still remains a final challenge to be addressed. The question is how much weight should the historically based reason carry in the final assessment of a present act? Does the terrible prehistory of the use of neurointerventions constitute a very heavy reason against the present use of such interventions? That is, does the history almost predetermine the answer to a present moral assessment? Or, is it rather the case that the historically generated reason constitutes a less significant reason that will only in a few cases tip the balance when the *pros* and *cons* of a present use of an intervention are almost equally strong? It might perhaps be felt that this challenge is a little unfair, insofar as it seems to reflect a much more general problem, namely, that it is often difficult to determine the respective weight of different reasons in moral evaluations. However, even if this is true, it does not change the fact that we need to know something about the weight of the historically generated reasons in order to be able to include these reasons in present decision-making. In the absence of an answer, the historical objection based on pessimist induction will be drained of practical guidance. The problem, of course, is that it is quite hard to see how this challenge should be answered.[16]

(E) The above challenges all arise if one holds that history inductively provides a reason against the current use of neurointerventions. However, even if we, for the sake of argument, assume that the challenges have been satisfactorily answered, we would still be faced with what constitutes the cardinal issue, namely whether the pessimist inductive inference in itself should

[15] There is some disagreement about the estimates of the total number of people who were subjected to lobotomy. As Swayze notes, 'just how many patients were actually operated on in the era of psychosurgery will probably never be known'. A survey of the number of patients who underwent lobotomies in England and Wales documented more than 10,000 single operations (and several hundreds who underwent more than one operation) (Swayze 1995, p. 511).

[16] Presumably, the weight of an historically generated moral reason is a function of the degree of wrongness of the acts constituting the base cases. However, beyond this overall point, it is hard to see how the weight of such a reason should be determined.

be regarded as a morally acceptable kind of reasoning. In my view, this is not the case. I admit that, in arguing to this effect, there is a risk of committing *petitio principii* by almost presupposing what should be established. If one really believes that the history inductively provides moral reasons, then one may not be persuaded by the arguments that follow, while, if one does not believe this to be so, then the arguments may appear much more persuasive. However, I believe that the only possible way of arguing to such an effect is by outlining a number of implications of the view which, hopefully, may affect the overall assessment of the plausibility of this inference. Here are two examples.

It is a well-known fact that technologies can be used in right or wrong ways (this is sometime referred to as the 'Janus-faces' of technology). For instance, a mobile phone can be used to save people by calling an ambulance and to kill people by detonating a bomb. Now, suppose that a person has been seriously hurt and I consider calling an ambulance. The right-making properties of this act of phone calling are obvious. The person will receive medical assistance, and she and her relatives will be saved serious problems. However, suppose it is the case that some years ago I used my mobile phone (on a sufficient number of occasions) to detonate bombs causing many people serious harm. These acts were clearly wrong. But does this fact make any difference whatsoever with regard to the rightness of presently calling the ambulance? I find this very hard to believe. Of course, it could be objected that it is not only my previous use of my phone that should be drawn into consideration, but also the way phones have been used by other people (which, presumably, has usually been unproblematic). However, if we hypothetically assume that my phone calls were the only ones to have ever occurred and that the inductive inference is valid, then it would be the case that my previous wrongdoing would produce a reason against calling an ambulance. Perhaps McTernan would object that, on her proposal, the fact that someone has been wronged is only one element of the determination of the proper inductive base cases; the other element is that scientific mistakes have occurred, which is not the case in the phone example. However, we could easily rephrase the phone example in such a way as to include scientific mistakes. For instance, we could imagine that I had previously used my phone to test the remote detonation of bombs in crowded areas, but that this procedure did not live up to the scientific standards for testing. It is hard to see that this slight change in the past use of the phone should in any way give more support to the idea that past wrongdoings constitute a reason against presently calling an ambulance and saving the person who has been hurt.[17] Moreover, it is worth noting that the example could just as well be turned around. Suppose that I consider detonating a bomb by making a phone call, but that in the past I have always used phone calls in decent ways, such as to call an ambulance when someone needed it. Detonating the bomb and thereby killing innocent people would obviously be wrong. But would the wrongness of this act in any way be mitigated by the rightness of my previous use of the phone? If a pessimist induction is plausible then it difficult to see why an optimist induction should not be equally valid. But it is very hard to believe that my previous proper use of my phone should provide any reason whatsoever in favour of my now making the call that would detonate the bomb.[18]

Another example to the same effect concerns the extent to which past wrongdoings would determine the wrongness of future acts if the pessimist induction is taken for granted. As mentioned, it would be absurd to subscribe to a strong version of the pessimist induction according to which past instances of wrongful acting, such as the use of a certain technology, would make all future instances of this act wrong. Arguably, such a principle would, due to the historical precedence, turn the current use of neurosurgery for therapeutic purposes into wrongful treatment.[19] Therefore, the view must be, as McTernan has underlined, that a history of wrongful use will provide *a reason* against present or future instances of this way of acting. However, even this modified interpretation of a pessimist induction may end up predetermining the moral assessment of all future instances of an act. Suppose, hypothetically again, that a new technology has been invented and that, even though there are both *pros* and *cons* for using this technology, it turns out that the reasons in favour will slightly outweigh the reasons against. Suppose, furthermore, that before the initiation of this long line of future rightful applications of the technology, I am now in a position where I can decide to use the technology (in a sufficient number of cases) either in a wrong or right manner. In that case, my present decision will predetermine the moral

[17] It could perhaps also be objected that the phone example fails because it is not the same act that is involved in the two cases. Rather than being two instances of the act of making a phone call, the example involves past acts of detonating a bomb and a present act of helping someone. However, this answer brings us back to the typing challenge outlined above; that is, the fact that it may not be clear what constitutes the proper description of an act.

[18] Obviously, this example could, if one insists that this is important, also be construed as a case involving a scientific mistake.

[19] As it has been put in a popular introduction to neurosurgery: 'It took humans no time to crack into a skull. Cracking the art of neurosurgery? That's a different story' (http://mentalfloss.com/article/70265/brief-history-brain-surgery).

assessment of all future cases. If I decide to use the technology in a wrongful way, this will, by induction, provide a reason against all future use. As long as this reason is sufficiently large to tip the balance between the *pros* and *cons*, all future cases in which the technology is used will be wrong. If instead, as the technology pioneer, I had decided not to use the technology wrongfully, then all future cases where it is used would become morally right. Admittedly, these are highly hypothetical circumstances. But it still seems hard to believe that one would in this way possess the power to predetermine the moral quality of all future applications of a technology.

The point of the above reasons (A) to (E) is not to suggest that they each constitute knock-down arguments against the idea of a pessimist induction in the outlined interpretation. I am unable show that answers cannot be given to some of the challenges. And someone who wholeheartedly believes that historical precedent provides reasons that should figure in a present set of right- or wrong-making properties of an act may not be persuaded by the final examples against the inductive inference. But, at the very least, I believe that the reasons provide a starting ground for considering whether a more plausible interpretation can be given to the present moral significance of the history of crime preventive use of neurointerventions.

Reinterpreting the Lesson from History

The dark history of how neurointerventions have been used in the treatment of offenders throughout the last century seems to leave us with a dilemma. On the one hand, it is very hard not to be affected by the history of how psychosurgery and drug treatment have historically caused tremendous harm to many people and to feel that historical cases must have some significance for modern discussion. On the other hand, the idea of interpreting the significance of history in terms of a pessimist induction seems to prompt several theoretical challenges that are not easily met.

Moreover, if we adopted the idea of a pessimist induction, then it seems that one could end up *not* paying sufficient attention to the past. The inductive inference—as we have seen—prescribes that, if the acts A_{-3}, A_{-2}, and A_{-1} were all wrong, then there is a reason against the present or future act A_0. But, presumably, the use of an inductive inference implies that, if it is possible to find a single or a few past instances of A-ing which were not wrong, then the inductive conclusion will be blocked. In other words, what adherents of a pessimist induction should be looking for is not only historical cases of how people have been maltreated in the past, but whether there exists one or a few occasions where this was not the case. Even though, as we have seen, many offenders have historically been treated in terrible ways by the use of neurointerventions, it could be the case that there is one instance in which such a treatment was not wrongly imposed. I will not here try to suggest such a historical case. Rather the point is, that if such a case exists, then apparently this will block the pessimist inductive approach to the present use of neurointerventions.[20] However, this seems to me to constitute too easy a way of dismissing the significance of history. Thus, even though an adherent of the pessimist induction, such as McTernan, no doubt subscribes to the inductive principle in order to emphasize the importance of history to our present considerations of neurointerventions, the implication of analysing history in terms of an inductive inference, might well be that the past loses moral significance if it is possible to find just a single or a few counterexamples. In my view, this also indicates that the pessimist induction is on the wrong track. Thus, the question is whether it is possible to reconcile the views that the history of neurointervention has something to offer in the modern discussion of this type of treatment, and that the pessimist induction does not constitute a plausible way of analysing the significance of the historical precedence?

In my view, the most plausible interpretation of a lesson from history approach to the present discussion of the use of neurointerventions for crime preventive purposes, is that the past can help us to sharpen our eyes when we consider the step from a theoretical moral discussion of neurointerventions to the practical application of this treatment. This step from theory to practice, in my view, is much larger than is often recognized in philosophical discussion, and the historical precedence can function as a valuable source of information to the various pitfalls that may exist when such a treatment scheme is considered being put into practice. Here are some examples.

[20] It might be objected that, if there are many cases in the past where X-ing was wrong, but also one or some cases in which X-ing was right (or at least not wrong), then the latter cases do not constitute counterexamples that block the inductive inference. Rather, what this shows is that there are both inductively warranted reasons *pro* and *con* for a present or future instance of X-ing. I will not here enter into a closer discussion of this possibility, except to remark that this view implies that historical reasons will end up playing a role in all present and future ways of acting (except brand-new types of acting), and that the theoretical challenges outlined in the previous section thereby become even more urgent concerns.

(1) *Do we know that a treatment involving neurointerventions will only be recommended by scientists when it has been sufficiently scientifically validated?* One of the most striking aspects of both the lobotomy era around the 1950s and the revival of psychosurgery in the 1960s and early 1970s is how these treatments were recommended by some scientists—underlining that the treatments were safe and efficient—while others critically-minded, held that these contentions were not backed up by any, or at least not sufficient, evidence.[21] In his critical examination of the scientific evidence concerning the use of psychosurgery, Sterling pointed out that: 'One must ask, therefore, whether psychosurgery has, after 40 years and 50.000 or more operations, developed any evaluative criteria. Are the claims of success supported by real evidence? Have the surgeons identified the contraindications and side effects that must accompany *any* therapy, let alone one that destroys part of the core of our beings?' (Sterling 1978, p. 150). In an evaluation of the scientific literature on psychosurgery from 1976 it was concluded that: 'The great majority of the psychosurgical literature has no scientific value and little validity. The possibility that a significant part of improvement seen after surgery can be attributed to biased selection of patients and 'placebo' effects cannot be ruled out' (Valenstein 1976, quoted by Sterling 1978, p. 151). Yet this treatment had been recommended and massively performed by surgeons. Sterling confirms how science can be biased when he reports one of his own experiences:

> At my own institution a patient who had received three lobotomies, which ultimately resulted in a hemiparesis, hydrocephalus, and severe epilepsy, was considered to have been 'helped' by the treatment. The patient, not institutionalized before the surgery, and then considered 'bright', now has an I.Q. of 61, no social life, no work, and *does* sit at home, passively, unresponsively. Human capacity for self-deception is almost boundless, which is, in part, why prospective, controlled studies are indispensable. (Sterling 1978, p. 151)

The conflict between the widespread use of a treatment and the lack of sufficient evidence concerning safety and efficacy was also found in various types of pharmacological treatment (see e.g. Spece 1972). An obvious reaction to such facts is to underline that science has undergone significant changes. As noted, medical experimentation is today subject to much stricter

[21] For an instructive review of the articles describing psychosurgery between 1935 and 1954 and, thus, of what was perceived as the reasons in favour of such operations and the increasing acknowledgement over this period of the side-effects, see Swayze (1995).

regulation. It is also a fact that most of the treatment in these periods did not live up to modern standards with regard to preceding clinical trials, acceptance from institutional review boards, and a more general codified ethics. But do regulations and institutional control constitute a sufficient safeguard against the mechanisms that drive scientific judgements and recommendations? And do we today even possess a sufficiently clear idea of what counts as *sufficient* evidence for a treatment to be put into practice in a group of individuals who do not even suffer from standard medical conditions?

(2) *Do we know that decision-makers will make decisions regarding the use of neurointerventions on the ground of proper ethical considerations and the best available scientific evidence?* Philosophical considerations, as this book also illustrates, often take place on the ground of stipulations which make it possible to consider overall principles governing a practice and the theoretical objections against it. Medical research may provide evidence of the effects and side-effects of a treatment. Criminological studies may provide evidence of how very different types of standard rehabilitation may help in reducing the risk of recidivism. However, in order to make the right decision on whether a treatment should be used, decision-makers will have to navigate in a field that requires interdisciplinary insight. Moreover, and more importantly, there is strong historical evidence for the fact that decisions at a political level may well not be attentive to principled ethical considerations and that they are quite often not academically informed. We will return to this important point shortly (Chap. 7).

(3) *Do we know that neurointerventions as crime preventive tools will be used only on those offenders for whom the treatment is appropriate in terms of efficacy and side-effects, and that they will be properly imposed?* The history of the use of neurointerventions provides numerous examples of how the administration of neurointerventions is vulnerable to various types of arbitrariness. This has clearly been the case in the broader therapeutic use of such interventions. For instance, Sterling mentions that, 'It is estimated in the 1970, that the average neurosurgeon performed only five or six operations per month . . . The pressure on these specialists to find adequate employment for their skills must be taken seriously, since it is known that the number of operations performed tends to be proportional, not to the demonstrated need for surgery, but rather to the number of surgeons available' (Sterling 1978, p. 150). Moreover, it is underlined that a single surgeon—cited above for his strong enthusiasm with regard to the possibilities of handling social problems through brain interventions—in this period himself conducted a

very high percentage of the total number of psychosurgeries.[22] If this is correct, then it clearly illustrates that the practical administration of a treatment may sometimes be heavily influenced by arbitrary factors such as institutional needs or perhaps the individual convictions of those who possess the requisite medical skills.

Highly dubious administration of neurointerventions in the treatment of prisoners is also manifoldly demonstrated in the historical periods considered above. There are numerous cases in which prisoners' rights *not* to be involved in experimentation or treatment were egregiously violated. For instance, commenting on a California Act supposed to protect this right at the beginning of the 1970s, Spece remarks that, 'While the Act gives persons involuntarily detained for treatment and evaluation the rights to refuse shock treatment and lobotomy, it also apparently gives the professional person in charge of the facility the power to deny those rights by merely 'rubber-stamping' a request for authorization to treat the inmate' (Spece 1972, p. 618). Indeed, it is not difficult to find cases in this and earlier periods in which prisoners have not consented to treatment. For instance, Spece reports that, in the above-mentioned case of drug therapy at Vacaville, at least five inmates were treated without giving their consent (Spece 1972, p. 635).[23] And even though consent today is no doubt taken much more seriously than in the 1960s and 1970s, can we honestly say we know that prisoners who are currently being treated always fully understand the information provided about the treatment?

Another noteworthy historical example concerns the misuse of treatment. In various cases, drugs that were supposed to be used as part of treatment programmes were given to prisoners for reasons having little or nothing to do with treatment. For instance, in an article from the mid-1970s, Edward Opton directed attention to how drugs were used as a way of punishing inmates. In an interview, a prisoner reported: 'If you speak out, say things they don't like, if you're a leader, you know—it's an unspoken threat: they'll put you on Prolixin'. Another prisoner said: 'They use Prolixin more for punitive action, than for medical purposes' (Opton 1974, p. 640). Opton's conclusion was that, while there were some constitutional limits placed on legal

[22] Sterling mentions that Dr. M. Hunter Brown accounted for over 40% of psychosurgery in the US in the early 1970s (Sterling 1978, pp. 127 and 137).

[23] In relation to the treatment programme at Atascadero, Michael Serber et al. (1975) note that 'All of these patients forcibly received succinylcholine without their approval—there was no signed consent' (p. 89).

punishment, corresponding limits did not exist with regard to the use of treatment. As he put it: 'Penal administrations turn, therefore, to therapy-as-punishment to carry out acts which, if named punishment, would be clearly illegal and immoral' (Opton 1974, p. 643). Along the same line, Serber et al. observed that: 'Far too frequently, however, behavior modification terminology as well as other psychiatric nomenclature has actually been no more than a pseudo-scientific rationalization for harsh and punitive measures of control within environments lacking resources to positively reinforce inmates' (1975, p. 87). Other commentators in this period emphasized how drugs were administered as tranquilizers in prisons, and underlined the major temptation of using drugs in this way instead of for physical methods of control (Sitnick 1975, p. 387). Does this temptation still exist?

The purpose of the outlining of all the above questions is not to suggest that they exhaust the number of issues that need to be addressed if one considers taking the step of implementing neurotechnological treatment in penal practice. Rather, they are supposed to indicate how the history of the treatment of inmates may function as a source we can draw upon in the attempt to get an overview of the various ways in which treatment can go wrong when put into practice. Furthermore, in my view, the questions, indicate the naivety of the contention that the answer as to whether neurointervention should be used in crime prevention is ultimately a question of whether scientists have succeeded in developing the technology. I agree with McTernan when she rejects the objection to the significance of the past, namely, that considerations of the dark history of neuroscience can be put to rest because we are now (or at least in the future will be) better at science. However, contrary to McTernan, I do not believe that the insufficiency of the objection has anything to do with reasons that follow inductively from the past. More simply, as the questions show, the point is that there are so many other things that matter in the ethical assessment and which could go wrong, even if, hypothetically, it is assumed that neuroscientists have succeeded in developing safe and efficient treatments. Thus, in this light, the history provides us with reasons to be cautious. As such, McTernan's pessimist induction could perhaps be regarded as useful, not as a principle that literally involves an inductive inference between past and present right- or wrong-making properties, but simply as a rule of thumb or a heuristic device, that can help us to pause before taking precipitate action and ensure that the mistakes of the past are not repeated today. It may be an overstatement, as the saying goes, that those who do not learn from the past are doomed to repeat it. But there is certainly a higher risk of doing so.

Conclusion

The history of how neurointerventions throughout the last century have been used as instruments of crime prevention, as noted, is slowly entering the modern discussion. However, despite a few and highly instructive papers on this topic, it is clear that there is room for more research on the precedents of the present resuscitated affair between neuroscience and criminal justice. As Francis X. Shen has rightly remarked, 'there is a rich, problematic, and underexplored history of neurolaw' (Shen 2016, p. 688). It is an interesting question as to why this history does not play a more prominent role in the contemporary academic discussion. For instance, it is remarkable how few references there are in the modern literature to the rather comprehensive debate—from which some arguments have been listed above—that took place in the 1970s. However, the aim here is not to contemplate the causes of academic amnesia.[24]

Rather, the purpose has been to consider what content, if any, could plausibly be given to the idea that history gives us a present reason to be cautious; or, more generally, whether a plausible interpretation can be given of what could be called a lesson from history objection to the present use of neurointerventions in criminal justice practice. What has been argued is that 1) the past—and indeed horrendous—use of neurointerventions does not per se provide a moral reason against present or future use of this sort of treatment. The idea that the significance of the past should be interpreted in terms of an inductive inference, according to which previous wrongdoings in themselves constitute a present wrong-making property, opened up several theoretical challenges which did not allow for an easy answer. Ironically, interpreting the significance of history on the ground of an inductive inference could, arguably, also have the implication that history was drained of significance; namely, in those cases where there exist counter-examples that block the inductive inference. Alternatively, it has been suggested that 2) The lesson from history should be given a more straightforward interpretation, namely, as underlining the importance of being conscious of and keeping in mind all the factors that play a role in the overall assessment of whether it is justified to use neurointerventions in criminal justice practice. Even if it is assumed that a neurotechnology has been developed in such a way as

[24] Shen suggests that reasons for academic amnesia may be due to shifting terminology over time, and the fact that it may be rather discouraging 'to look backward to see how one's professional field failed at previous endeavors' (Shen 2016, p. 688).

to make it possible to reduce recidivism without causing too serious side-effects, there would still be so many ethically relevant pitfalls in the route from actual scientific recommendations, to political decision-making, and practical implementation, that need to be addressed. The historical lessons can help us to ask the right questions.

But if this is how the significance of history should be interpreted, then we are immediately faced with a new question, namely, what are the answers to questions such as those posed in the previous section? Obviously, there is no simple answer to this. It all depends upon what type of neurointerventions we are considering, for what purposes, and, not least, in which institutional and societal contexts. Unfortunately, I cannot thoroughly engage in such considerations here. However, more modestly, what I will do in the following chapter is to address one important aspect of a present context which, I suggest, gives reasons for serious concern.

7

Neurointerventions and Criminal Justice Practice

A methodological distinction which, in recent years, has attracted increasing attention in normative theorizing, is the one between ideal and non-ideal theory. What has driven the invocation of this distinction, at least partly, is the frustration of mainly political philosophers, but also other normative theorists, with the fact that theories addressing highly important societal issues have often had very little, if any, influence on real-world political decision-making. In my view, this is the case with regard to several (not all) areas of ethics, and also to the one that is relevant for many of the considerations in this book, namely, the ethics of punishment.

This may sound surprising. After all, penal theorists have often not been concerned with purely hypothetical topics but have addressed what seems to constitute highly pertinent issues such as the workings of the criminal justice system and, more precisely, questions as to why and how offenders should be punished. All questions that are close to real-world practice. However, there has been a tendency to engage in such considerations almost as if it were possible to design and build up a penal system from the start; that is, as if such a system did not already exist. Little *theoretical* attention has been directed to the facts that criminal justice systems may not be easy to change; that it is far from obvious that theoretical considerations are regarded as at all welcome by those who possess the power to make decisions; and, not least, that criminal justice practice may currently deviate significantly from what ethical theories presuppose and prescribe. Unfortunately, and despite the fact that I regard this as a highly important and somewhat ignored issue in modern penal theory, I cannot here engage in more general considerations of the theoretical implications of conducting penal theory under non-ideal circumstances. However, what I will do in this chapter is to address the question as to how the way criminal justice systems currently function and deal with offenders should influence our considerations on the possible use of neurointerventions in criminal justice practice.

Neurointerventions, Crime, and Punishment. Jesper Ryberg, Oxford University Press (2020). © Oxford University Press.
DOI: 10.1093/oso/9780190846428.001.0001

From the outset, there are several things that complicate this discussion. First, criminal justice practice varies among different countries. In the following, I will outline some of the characteristics of criminal justice practice in the United States. But it is a well-known fact that this practice differs in several ways from how other Western countries deal with criminal offending. However, despite the fact that punishment practice in the United States since the early 1970s has undergone changes that are unparalleled in other countries, there are nevertheless some tendencies which are shared across several Western countries and which are relevant in considering the possible use of neurointerventions. Second, if we are considering types of neurointervention that may be fully developed in the future, then we cannot be certain that criminal justice practice has not undergone changes. Even though criminal justice systems in general rarely undergo very rapid changes, one should be wary with regard to simply extrapolating present trends into a more distant future. Third, obviously one cannot know for certain how precisely a new technology would be implemented and used within the criminal justice system, even if one were to know how this system is currently operating.

What all this shows, of course, is that there are reasons to be cautious not to draw too strong conclusions when one engages in considerations of what to expect with regard to how crime preventive neurointerventions may be used in practice. However, as we have seen in the previous chapter, practical use of neurotechnology may be encumbered with various pitfalls which are easily overlooked if one stays within the spheres of ideal theorizing. Thus, the time has come to consider the risks that may be involved in introducing crime preventive neurointervention within what must be characterized as conspicuously non-ideal circumstances. As will be argued, even if one imagines that a neurointervention has been developed which in some cases will have a crime preventive effect without causing too severe side-effects, one may still be well-advised not to introduce this treatment under circumstances that equal those currently existing in criminal justice practice.

In order to sustain this conclusion, the remainder of this chapter will be structured as follows. We will start by directing attention to some of the significant characteristics of current penal practice in the United States; that is, more precisely, the facts that establish what many commentators regard as the that lack of rationality in the way offenders are being punished. It will be argued that if these characteristics constitute the framework within which neurointerventions are introduced, then there are strong reasons to fear that such a treatment will be administered in a morally wrongful manner. This will be shown by reassessing many of the questions that have been addressed in the previous chapters in the light of a striking non-ideal feature of penal practice, namely, that many offenders are being punished too severely. It is argued that the tendency to severely over-punish offenders has significant implications with regard to how neurointerventions may be used in the prevention of crime. Finally, I will sum up the considerations of this chapter and provide a picture of what I regard as the main overall conclusions that can be drawn from the previous chapters. As will become clear, I believe the distinction between ideal theory and non-ideal real-life circumstances to be of crucial importance for the approach that should be adopted to the use of neurointerventions as a crime preventive instrument.

Traits of Criminal Justice Practice

One of the most striking features characterizing criminal justice practice in the United States is the huge number of offenders who have been placed behind bars. The history of how this comprehensive use of incarceration has come about over the last couple of decades has often been told. Until the 1970s, imprisonment rates had been fairly stable in the United States for most of the 20th century and no particular trends were apparent when compared to many other countries (Tonry 2009). However, at the beginning of the 1970s, the imprisonment rate started increasing. Despite the fact that crime rates declined from the early 1990s, prison growth accelerated. The imprisonment rate continued rising until the end of the first decade of the new millennium (reaching a peak at more than 750 inmates per 100,000 population). Even though there have been some fluctuations over the last couple of years, the overall picture is that the imprisonment rate in the United States has quintupled over a few decades and has reached levels that are not only unprecedented in US history but also in the history of liberal democracy (Garland 2001).

The widespread view on this development, and on the current status of the use of incarceration, is that something has gone entirely wrong. For instance, influential commentators talk of the epidemic of mass incarceration from which the United States is suffering and have labelled this a 'tragedy and a national embarrassment' (Tonry 2016, p. 441; see also Husak 2019). However, with regard to the question as to precisely *how* this development should be explained there is less agreement, except when it comes to the unsurprising view that many factors have contributed (see e.g. Garland 2001; Lacey 2008; Pharff 2017; Tonry 2004, 2009; Zimring 2001). At the most trivial level, an increase in the prison population is the result of the fact that more offenders are being incarcerated and/or that offenders are being incarcerated for longer periods. What constitutes the relative contribution of each of these two causes to the present state of mass incarceration is not something that we need to address here (see e.g. Pfaff 2017). What matters for the present purpose is the fact that offenders are being punished more severely and, in particular, that they are being punished *too* severely. This is the case not only in the United States but also in many other Western countries.[1]

That many offenders receive excessive punishment is of course a normative judgement and, hence, a judgement that cannot be sustained without reference to ethical considerations. As we have seen, there exists no agreement with regard to what constitutes the most plausible ethical approach to punishment. At one end of the scale, there are versions of retributivist theory emphasizing desert and desert-based proportionality constraints as constituting the plausible approach to punishment. In these views, considerations of consequences play only a very modest role (and in a few extreme cases perhaps no role at all). At the other end of the scale, we find purely consequentialist approaches, such as the standard utilitarian theory. In between these two poles, there is a variety of theories which give varying significance to desert constraints and consequences. In fact, the picture is even more multifaceted. For instance, as has also been indicated in the previous discussion, theorists may agree on the overall question as to whether there are particular ethical constraints on the proper use of punishment, while providing very different theoretical expositions and justifications of such constraints (see Chap. 4). However, despite all these variations at the theoretical level, there exists a striking consensus with regard to the view that many offenders are being punished much too severely.

There are two ways in which offenders can be over-punished. The first relates to the problem of overcriminalization. Some theorists believe that there are, currently, several ways of acting that should not have been criminalized in the first place. For instance, Dough Husak, who has comprehensively

[1] For considerations on the changes in the use of imprisonment in England and Wales over the last two decades, see e.g. Roberts and Ashworth (2016). As the authors note, the 'principal causes of the increase over the 20-year period [1993–2015] were harsher sentencing, including higher custody rates and longer prison sentences' (2016, p. 313). For a more general discussion of sentencing trends in England, see e.g. Tonry (2004). For an excellent overview of penal policy developments in five Western countries, see Roberts et al. (2003).

considered this issue, describes overcriminalization as 'the most pressing problem with the criminal law today' (Husak 2008, p. 3). But if there are currently too many laws on the books then it follows that the criminal sanction is being overused; that is, some people are being punished even when they should not have been punished at all. In this way, an implication of overcriminalization is that it produces over-punishment (see also Ryberg 2018). The other, and perhaps more obvious source of over-punishment, is that, while offenders have committed acts that have rightly been criminalized, these are being punitively reacted to in too severe a manner.

That many punishments are over severe, that is, that many offenders spend too much time in prison, is a view to which many retributivists—on the ground of different desert-theoretical positions—have in recent times subscribed. To give a few examples, Richard Singer has underlined that it is a misconception to think of the desert model as a derivative of a 'throw away the key' approach to punishment. He has suggested that, in contrast to what is practiced in many jurisdictions, confinement should be reserved only for the most serious crimes and, even then, the duration should be relatively short (Singer 1979, p. 44). Along the same lines, another retributivist, Jeffrie Murphy, holds that if desert theory were to be followed consistently one would punish less and in more decent ways than one actually does (Murphy 1979, p. 230). And many more recent retributivists hold corresponding views. For instance, Andreas von Hirsch has suggested models for scaling down punishment levels and has even contended that terms of imprisonment should seldom exceed five years (see von Hirsch 1993; von Hirsch and Ashworth 2005).

Corresponding views have been suggested from a consequentialist perspective. Studies of deterrence and incapacitation have shown that there has been no gain in terms of crime prevention by locking offenders up for longer periods and, consequently, that there are reasons to seek to reduce penal levels. In a recent discussion of the effects of sanctions in the US penal system, Tonry summarizes his review of studies on crime prevention in the following way: 'In 2017, it is not controversial to assert that the crime prevention effects of mass incarceration have been much less than many people supposed or hoped, that there is little or no reason to believe that harsher punishments have greater deterrent effects than milder punishments, that incapacitating people by locking them up for lengthy periods is an ineffective crime prevention strategy, or that the experience of imprisonment makes many offenders more, not less, likely to commit crimes later in their lives' and, furthermore, that: 'The implications of the literatures on deterrence and incapacitation are straightforward: few convicted offenders should be sent to prison and for shorter times' (Tonry 2016, pp. 453 and 459).[2]

What has contributed to the use of too harsh prison terms in the United States are initiatives such as mandatory minimum sentence laws, life without parole laws (so-called LWOPs), and truth-in-sentencing laws. However, in addition to these initiatives, there are other laws that have led to longer sentences, which are relevant in relation to our discussion of the use of neurointerventions, and which also conspicuously demonstrate the discrepancy between penal theory and practice. What I have in mind is the role of criminal history. That an offender's prior criminal record can lead to sentencing enhancements is a feature in all contemporary sentencing regimes in Western countries (see Roberts 2008). In the United States, the most controversial way of taking criminal history into account is the Californian strike law of the mid-1990s. However, there are differences between the role played by criminal history in the various states. For instance, in states using two-dimensional sentencing grids with criminal history constituting one of the dimensions, the ratio of the recommended sentence length in the highest criminal history category relative to the sentence in the lowest criminal history category is on average 4.7 in Minnesota, while it is around 10:1 in Washington, and 14:1 in Kansas (Roberts and Frase 2019). However, the general picture is that the influence of criminal history has contributed to mass incarceration and, more importantly for the present purpose, that it implies that recidivists often spend a long time in prison. Moreover, in some cases involving less serious crimes, taking into account a prior criminal record implies that offenders, who would otherwise have been placed on probation, end up in prison. But, is the fact that criminal history is attributed the role of key component in sentencing based on a solid theoretical ground? As indicated, it is generally agreed by commentators, that the answer is in the negative.

From a retributivist perspective, the views on the moral significance of a prior criminal record are divided between a number of different positions (Roberts and von Hirsch 2010; Tamburrini and Ryberg 2012). Roughly, the theoretical picture is the following. Some theorists subscribe to a flat-rate position (Fletcher 1978; see also Tonry 2010). According to this view, the offender's criminal history has no bearing on the sentence which he or she

[2] Tonry also holds that: 'Imprisonment of most people over 35, from a crime prevention perspective, is a waste of money' (2016, p. 459).

deserves. No matter whether the offender is a first-timer or has a comprehensive criminal record, this does not affect the desert debt of the present crime. Neither the harm caused by this crime, nor the culpability of the offender, are influenced by the prior criminal history. Most writers, however, seem to believe that a criminal record should play some role. Amongst those who have defended this view, some theorists subscribe to the idea that the absence of prior convictions (or a minor criminal record) should serve as a mitigating factor (von Hirsch 2010; von Hirsch and Ashworth 2005; but also Ryberg 2010). On this view, usually referred to as the *theory of progressive loss of mitigation*, an offender should receive a punishment that is less serious than that which is deserved out of considerations of harm and culpability. As the offender reoffends, this initial mitigation is gradually lost. Thus, following this idea, an offender will never deserve a punishment which is more severe than that warranted on the ground of considerations of harm and culpability in relation to the present crime but will, in some initial cases, be allotted a less serious punishment. The alternative view is to hold that criminal history should play a role as an aggravating factor. In this account, the deserved punishment increases as an offender acquires more and more prior convictions (Dagger 2012; Davis 2012). However, it is fair to say that attempts at justifying this view are highly controversial and, more importantly for the present purpose, none of the attempts at developing this view have advocated aggravations of a magnitude that is seen in penal practice (see e.g. Davis 2012). Thus, as Julian Roberts and Richard Frase have put it, 'retributivist punishment theory fails to supply a strong justification for prior record enhancements, especially of the kind applied in US guidelines systems (2019, p. 6).

But what if we leave retributive theory and turn instead to a consequentialist approach? Is it not precisely the case that prior convictions constitute a standard predictor of recidivism and that the imposition of harsher punishments on repeat offenders is, therefore, justified as a way of curbing future re-offending? Even if such a justification sounds tempting, it is in a several respects flawed. As Roberts and Frase have underlined, a crime preventive justification presupposes: 1) that prior offending reliably predicts future offending; 2) that the higher risk of re-offending is serious enough to justify harsher punishments; and 3) that prior record sentencing enhancements effectively manage to prevent further offending (Roberts and Frase 2019, pp. 7 and 8). However, the problem is that evidence does not support these presuppositions. In their review of the empirical research, Roberts and Frase ultimately conclude:

> We can summarise the principled ways in which criminal history enhancement across the US diverges from empirical risk-related patterns with the following points. First, the structure of prior record enhancements, imposing steadily-increasing penalties for each increment in criminal history, is inconsistent with what we know about how recidivism rates increase with increases in the number of prior convictions. Second, guidelines include prior crimes and other record-based variables with little or no predictive power in terms of re-offending (due to the remoteness in time or a lengthy gap between previous and current offending) . . . Third, the enhancements increase the severity of imposed punishments in ways and magnitudes which are inconsistent with current knowledge of the effectiveness of different penal sanctions. These include imprisoning an offender to reflect his prior convictions when imprisonment results in higher, not lower rates of re-offending, and imposing longer prison terms when this in unlikely to reduce recidivism. (2019, p. 13)

The picture is that current prior record enhancement lacks justification both from a retributivist and a consequentialist perspective. In this way the significance of prior criminal history in sentencing constitutes just a special example of the above-mentioned problem that many offenders receive punishments that are too severe.

Another and more general way of putting the above points is to say that, if by a penal system being rational we mean, first, that it is based on what constitutes the most plausible overall theory of justification and, second, that it draws on what constitutes the available scientific evidence to ensure that the goals specified in the justificatory theory are reached, then a criminal justice system operating in the manner described above is very much not a rational system (see also Tonry 2001). This is what characterizes the criminal justice in the United States and, I believe, in many other Western countries.[3] Even though much more can of course be said on the way the criminal justice system works in the United States and in other Western countries, sufficient

[3] In the present context, it will be much too comprehensive a task to provide an overview of trends in sentencing and criminal policy in Western countries. However, it should be mentioned, that there are several countries that have done very little to specify a basic rationale for sentencing. Conversely, some countries specify a list of both forward-looking and backward-looking justifications—what Ashworth once referred to a 'smorgasbord' approach—without realizing that these goals may lead to inconsistent prescriptions. As for the tendency to pass new legislation without drawing on scientific evidence, this can be found in many countries. For instance, this tendency has also become dominant in those which are usually regarded as not being 'tough on crime', such as the Scandinavian countries (see e.g. Lappi-Seppälä 2019).

has been said to indicate the widely accepted view, namely, that there currently exists a significant discrepancy between ideal penal theory and penal practice.

Neurointerventions Under Non-ideal Circumstances

Suppose that the somewhat rough picture provided above on the development and current status of part of criminal justice practice is accurate. Why would this be of any relevance for the discussion of the possible use of neurointerventions in crime prevention? The simple answer is that what the lesson from history, as interpreted above, has taught is the importance of asking the right questions. In other words, the question that arises now is what would happen if neurointerventions were to be introduced within a criminal justice system as it is actually operating? Admittedly, as initially indicated, this is a speculative question and therefore not one that admits of cast-iron conclusions. However, if we wish to avoid the naivety or reality-blindness that may be associated with ethical considerations when they are based on purely ideal presuppositions and seek instead to identify some of the risks and pitfalls that exist when a technology is put into practice, then this is the type of question that we will have to address.

Therefore, what I intend to do here is to re-examine some of the questions that have been addressed in previous chapters, but now from the perspective of a non-ideal penal context, the contours of which have just been outlined. As part of the ethical discussion of the various objections raised against the use of neurointerventions, what I have argued is that many of these objections were non-conclusive in the sense that, even if the moral premises were taken for granted, they did not suffice to establish that neurointerventions could not be used in a morally acceptable way. Some objections only managed to show that some types of neurointervention were wrong. Other objections failed to recognize that the identified moral problems could be accounted for in the way neurointerventions are used. However, the question now is what follows if we leave such principled considerations and start reconsidering the objections in the light of a non-ideal criminal justice context.

(A) *Coercive treatment*. One way of applying neurointerventions in the criminal justice system is to offer treatment to offenders, but to do so by making the duration of the prison term conditional on whether the offer is accepted. As we have seen (Chap. 2), this idea of tying together the period of incarceration and the treatment gives rise to the question whether an offender, under such circumstances, is at all free to make a choice. That is, whether the standard voluntariness condition of consent is satisfied. There has sometimes been a tendency to hold that if much is a stake, such as is clearly the case if one is faced with the possibility of staying for a longer or shorter period in prison, then the choice is not voluntary. However, as we have seen, this answer is premature. People make various sorts of decisions under tough choosing conditions (e.g. decisions concerning therapeutic medical treatment) where it is not plausible to hold that the choice is involuntary, and that consent is impossible. Attempts at explaining the involuntariness of a choice, between either a fixed prison term or a reduced period of imprisonment combined with treatment, in terms of some sort of meltdown in the person's choosing capacity did not seem plausible.

However, it might still be held that, even though a person's capacity for choosing is intact, the person may nevertheless still be placed under circumstances where it is reasonable to hold that a decision is made under coercion. This is precisely what happens when a person is threatened to act in a certain way. Thus, this opened up the discussion as to whether a proposed penal reduction accompanied by treatment should be regarded as a genuine offer, or whether it should be interpreted as a sort of threat. As we have seen, there is no consensus in the discussion as to how one should distinguish between offers and threats. Most theorists agree that the distinction depends on what constitutes the baseline of a proposal. An offer proposes to make someone better off relative to some baseline, whereas a threat proposes to make someone worse-off relative to some baseline. However, disagreement exists when it comes to the question as to what should be regarded as the proper baseline for the comparison. While, as we have also seen, some theorists believe that the baseline should be interpreted descriptively or predictively—that is, as concerning what would be the case if the proposal were not accepted—others believe it should be interpreted normatively, that is, as concerning what should be the case if the proposal were not accepted. Why would one hold the latter view? Nozick originally illustrated the idea behind a normative baseline by giving the following example: suppose that usually a slave owner beats his slave each morning, for no reason connected with the slave's behaviour. Today he says to his slave, 'Tomorrow I will not beat you if and only if you now do A' (Nozick 1969, p. 450). The point of this bizarre example is to show that if one relies on a purely descriptive or predictive baseline, then it seems that the slave owner's contention is an offer (the slave is not worse-off than he would be if the communication had not taken place; he would be beaten anyway). However, as Nozick underlines, it is tempting to view this as a threat. But this indicates that one should adopt a normative baseline (the slave is worse-off relative to how he should be treated: he should not be beaten every morning). Now, what I argued in Chapter 2 was that no matter whether one subscribes to a descriptive/predictive baseline or a normative baseline, this will not affect the fact that a proposed neurointervention and penal reduction would constitute an offer. Predictively, the offender would not be worse-off if the proposal is not accepted (he will receive the punishment he would have had anyway). And, normatively, he would not be worse-off because he should be punished. However, as was also underlined, the latter contention presupposes that the punishment is in fact justified. It is precisely at this point, it becomes relevant to consider circumstances of real-world penal practice.

If an offender is kept behind bars until he at some point accepts a proposed treatment, then, given the assumption that such continued imprisonment is unjustified, this would constitute an instance of coercion. Historically, this is what we have seen when treatment has been administered within a system of indeterminate sentencing.[4] However, even if we consider the existing determinate sentencing practice, the proposal of a reduced punishment would still amount to an instance of coercion, if the baseline punishment is not morally justifiable. Therefore, if it is the case that offenders are being punished too severely, then, from a normative base-line perspective, the offer of penal reductions might not constitute a genuine offer but an instance of coercion.

The consent-based objection against offering penal reductions in return for participation in some sort of neurorehabilitation could, as was also argued, be challenged on the ground that it is not so clear that crime preventive treatment requires the kind of consent which is usually required in standard therapeutic treatment. As we will see shortly, the arguments in support of this claim may also be challenged if the question is reconsidered under non-ideal circumstances. However, for the present, what we have seen is that if we move from ideal considerations of whether proposed neurointerventions could be justified to a real-life context, such as the present, in which offenders are being over-punished, then this might generate reasons against such

[4] As Serber et al. underline in their criticism of treatment under conditions involving indeterminate sentences: 'patients and prisoners are frequently under pressure to cooperate or face extended incarceration. Patients committed until they are 'cured' may in fact be facing a life sentence if they are judged uncooperative or unmotivated for treatment' (Serber et al. 1975, p. 88).

treatment proposals.[5] But this is only a first indication of the possible significance of considering the use of neurointerventions under non-ideal circumstances. As we will now see, the implications may be more far-reaching.

(B) *Compulsory treatment*. The use of neurointerventions as a treatment offered to offenders is controversial, but much less so than the possibility of using such interventions as part of compulsory treatment schemes. From a historical perspective, it is indeed understandable that the idea of compulsory treatment will be met with strong opposition. As we have seen in the previous chapter, the amount of harm that has been imposed on offenders in the name of treatment is tremendous. But, even if one ignores the historical perspective, it is clear that there are obvious reasons against such treatment. A forced neurointervention may at best be highly unpleasant and at worst egregiously traumatic. Thus, as we have seen, there is an immediate ground for regarding such interventions as *pro tanto* wrong. However, this is also the case with regard to the imposition of punishment. Punitive treatment involves the imposition of harm and constitutes an obvious example of compulsory treatment. Thus, the theoretical question that arises is whether it is possible to consistently hold, as most theorists do, that there are reasons to reject the former type of compulsion, while subscribing to the view that the latter type is acceptable (in fact, morally required). Are there reasons to believe that the use of neurointerventions should be considered morally on a par with punishment, in the sense that such treatment is *pro tanto* wrong but, in some cases, acceptable or morally required, all things considered? Or, should such interventions, all things considered, be rejected? I shall not recapitulate further elements of the previous discussion. Rather, the point here is to direct attention to ways of using neurointerventions for crime preventive purposes that would be morally wrong, but which may nevertheless be what one would experience if such treatment were to be put into practice within a system deviating significantly from what is ideally desirable.

An obvious way in which the use of neurointerventions would be wrong is if such treatment were forced upon an offender, and had undesirable side-effects, but *no* desirable effects in terms of crime prevention. This would be the case, if the treatment were to be imposed on the wrong person, i.e.

[5] It should be underlined that the purpose here is not to defend a normative baseline, nor to consider the strength of the moral reason against neurointerventions that are coercively proposed. As we have seen earlier, there could be penal theoretical reasons in favour of proposing penal reduction within a system that is over-punishing offenders. If the treatment is accepted, this would help in making the punishment morally more acceptable (e.g. less disproportionate). However, the point here is simply to outline these possible implications of over-punishment.

someone who should not have been treated. Basically, this would amount to an instance of harm infliction which, in the absence of countervailing desirable reasons, would be wrong from any ethical perspective.[6] What I have assumed so far—and what constituted the background for engaging in the discussion of the use of neurointervention in the first place—is, therefore, that such treatment could be administered without too serious side-effects *and* that it would have a crime preventive effect by reducing the offender's propensity for crime. In a philosophical discussion, such an assumption is easy to make. In practice, it is encumbered with challenges.

The obvious problem is that in order not to end up imposing harmful treatment on someone for no reason, one would have to be able to identify those offenders who are likely to reoffend. However, this is notoriously a problematic issue. Considerations of risk currently play an important role in the way criminal justice systems deal with criminal offenders. This is the case in the United States and in the jurisdictions of many other countries (see Keijser et al. 2019). However, many studies have shown that the error rates of risk assessments are substantial. In the mid-1970s, Norval Morris held that predictions of future violence were wrong as much as two-thirds of the time (Morris 1974; see also Tonry 2019). Today, the methods of risk assessment have changed. They are no longer based solely on clinical judgements by doctors and psychologists but draw on various sorts of actuarial analysis (see Ginneken 2019; Tonry 2019). However, many studies show that there are serious problems in terms of false positives (when reoffending is predicted but would not have occurred). For instance, an influential meta-analysis of violence prediction concluded that, on average, positive predictions are correct 42 per cent of the time (Fazel et al. 2012). Other analyses have reached corresponding results (see e.g. Tonry 2019). It has been underlined that group data provide poor instruments when it comes to prediction in individual cases. For instance, Seena Fazel et al. hold that 'risk assessment tools in their current form can only be used to roughly classify individuals at the group level, and not to safely determine criminal prognosis in an individual case' (Fazel et al. 2012, p. 5). Moreover, another serious problem related to the use of risk

[6] The point is not to hold that any system that happens to administer treatment on someone who should not have been treated would always be acting wrongly. Analogously, all penal theorists believe that a punishment system can be justified even if it is sometimes the case that innocents are being punished. And from a consequentialist outlook, it could be the case that a treatment system which in some cases misidentifies those who should be treated, would nevertheless produce the best overall result. However, as we will see below, the mechanisms that could lead to misidentification leave little hope that the overall result would be morally preferable.

assessment is that such instruments adversely affect minority offenders (see e.g. Tonry 2019).

But if there exist such problems when it comes to estimates of the risk that an individual offender will in fact recidivate, then what does this mean for the use of neurointerventions? Does it imply that as long as prediction instruments are not more reliable and one is not able to totally eliminate the existence of adverse side-effects (which is highly unlikely), then one will have to be highly reluctant to use such interventions or, at the very least, make sure that the risk of imposing pointless harm on offenders is given sufficient weight in an everything-considered judgement? Within an ideal or rational system, the answer to this question would be in the affirmative. However, if the answer is given within a framework such as the existing penal system, then it may take a very different form. If it is already the case that risk-based sentencing is being applied to an extent that clearly exceeds what is warranted on the ground of the reliability of the predictive instruments, then one might well fear that the same would be the case in the use of neurointerventions. Obviously, one cannot strictly infer from the existence of the over-use of standard criminal sanctions to the over-use of neurointerventions if these were introduced within the criminal system. But if one looks at the mechanisms that have contributed to the contemporary state of over-punishment, then the inference no longer seems premature.[7]

For instance, one of the mechanisms that have driven the excessive use of punishment is the so-called false-positive problem. In terms of risk-based sentencing, there exist two types of mistake. One mistake is not to send someone to prison—or to fail to keep a person incarcerated for a longer period—who commits a crime that could have been prevented had he been behind bars. This is a false-negative result (the person was incorrectly assumed not to constitute a risk). Another mistake consists in incarcerating someone who is not a genuine risk, that is, who would not have committed a crime had he been free. This, as we have just seen, is a false-positive error (the person is incorrectly assumed to constitute a risk). In theory, it is a standard position to hold that the criminal justice systems should be much more concerned about false positives than false negatives.[8] However, in real life the

[7] As earlier noted, it might be held that the administration of a harmful neurointervention on someone who is a false positive need not be a problem if the intervention is imposed as a punishment (see Chap. 4). However, as we will see shortly, the use of neurointerventions as punishment under non-ideal circumstances does also give rise to serious concern.

[8] For instance, this is standardly expressed in terms of the Blackstone ratio: that it is better that ten guilty go free than that one innocent be convicted.

picture may be quite the opposite. And it is easy to see why. As John F. Pfaff has appositely phrased the problem: 'The costs of a false negative are immediate and salient. Those of a false positive are nearly invisible and abstract. In the case of a false negative, there is an identifiable offender and an identifiable victim of the resulting crime, as well as an identifiable official at whom voters can direct their anger. The media and political opponents can ask, "Why did you release Bob? Why did you expose Mary—here's a picture of her with her two cute children—to the risk of victimization?"' (Pfaff 2017, p. 168). In other words, the emotional reactions to the two types of mistake are usually highly asymmetrical and those who can be held responsible for the mistakes may well respond accordingly. In a system, such as the one in the United States, where there are elected prosecutors and judges, this may influence penal decisions. In Pfaff's words, Judges and prosecutors may be 'inclined to over-punish to avoid the risk of a false negative blowing up a reelection campaign' (Pfaff 2017, p. 169). And politicians may be affected along the same lines.[9]

However, if the false-positive problem constitutes one of the causes of over-punishment, then it is not difficult to imagine that the same mechanism would also tend to produce overreactions in the use of neurointerventions, if such treatment were to be implemented in the actual criminal justice system. There would be a strong reason to seek to avoid false negatives, that is, cases where a treatment is not administered because an offender is incorrectly identified as not being a risk, but a comparatively less strong reason to try to avoid cases where the administration of a treatment is inefficient because the offender would not have reoffended anyway. This is the case with regard to the question whether a neurointervention should be administered in the first place. The false positive problem may push decisions in the direction of imposing treatments on more offenders than is necessary. However, this is only one side of the problem.

The other side concerns the reluctance to terminate treatment which has been initiated but which is no longer required in terms of crime prevention. The risk of ending up with a false negative outcome may provide a strong incentive to keep on treating someone even when he or she no longer

[9] Tonry contends that: 'The American system ties criminal justice policies and decisions in individual cases to election returns and to officials' personal and political self-interest. Rather than buffer policy and decisions about individuals' fates from emotional overreactions, moral panics, and changing sensibilities, American institutional arrangements are almost designed to assure their influence' (Tonry 2004, p. 19).

constitutes a risk. In public debate, it is certainly not uncommon to hear complaints along the lines: 'How can it possibly be the case that this person has been released from prison? We know he has a criminal background and now he has committed a new violent crime. Why is he out? Who is responsible?' (see also Pfaff 2017, p. 168). It is very easy to imagine precisely the same complaint presented in relation to absence of treatment. If there is a neurointervention which has been proven to have a crime preventive effect, then 'Why has this treatment been terminated? Who is responsible for the fact that the offender has recidivated?'. In fact, the disinclination to stop a treatment once it has been initiated may be even stronger than the disinclination to release offenders from prison. After all, even if the false-positive problem gives an incentive to keep offenders behind bars for longer periods than is warranted out of considerations of risk, considerations of retributive fairness may also play a role in relation to a punishment. Most people believe that there are limits to how long a time one should spend in prison for having committed a certain crime, at least if the crime is not very serious. However, corresponding fairness considerations are unlikely to have any influence on considerations of treatment, which is not imposed because an offender has committed a crime, but in order to prevent future crimes. Thus, the fact that one cannot even resort to fairness considerations may imply that a criminal justice system is even more vulnerable to the incentives that follow from the problem of false-positives. Historically, as we have seen in the previous chapter, this is precisely what occurred in periods in which various sorts of treatment were used in an almost unconstrained manner. As part of his historical overview, Rakoff notes that 'the law has often been too quick to accept, and too slow to abandon, the 'accepted' brain science of the moment' (Rakoff 2016, p. 31). Somewhat along the same lines, there is reason to fear that the implementation of neurointerventions—even if such treatment would in some cases have a crime preventive effect—within a criminal justice system dominated by serious problems of over-punishment, would also, for some of the same reasons that cause the overly use of criminal sanctions, result in a pointless imposition of harmful treatment on offenders.[10]

[10] What I have done here is only to direct attention to one of the mechanisms that have driven the use of harsher punishments and to suggest that this mechanism may well also lead to over-reactions in the use of treatment. Obviously, there are many other mechanisms that have contributed to the use of excessive punishments (see e.g. Tonry 2004, 2016; Pfaff 2017). In my view, many of the other mechanisms that have been suggested give further reason to fear overreactions if neurointerventions were to be introduced within the existing penal system.

It might be objected that it is difficult to honestly believe that one would, within the framework of the existing criminal justice system, force someone to undergo harmful and pointless treatment. However, as indicated, this objection seems naïve. Keeping someone behind bars is a very serious and harmful way of treating people, which not only affects the life of the imprisoned but may also cause tremendous collateral damages. Thus, if offenders can be placed behind bars in ways that are not justified, then there is no reason to suppose that unjustified treatment would not also be administered on this group of people (which is what history has abundantly demonstrated), and that some of the same forces that drive over-punishment may not also fuel tendencies for over-treatment. However, the use of neurointerventions on someone who should not have been treated thus, is only one way in which compulsory treatment might go wrong in current penal practice.[11] Here is another.

In considering whether a criminal sanction is morally justified there are of course many things that need to be taken into account. What we have focussed on in the present discussion are the harsh terms of imprisonment. Depending on the theoretical perspective the current use of such terms may, as we have seen, be regarded are morally wrong because, in comparison to shorter terms, they are of no benefit in regard to crime prevention (they can even be counterproductive) or, because they are gravely disproportionate. However, another question that should be considered in an ethical assessment is whether there exist preferable alternatives to a sanction. From a consequentialist point of view, it is obvious that the use of a particular sanction would be wrong if there exists an alternative type of sanction that *ceteris paribus* would cause less suffering or be preferable in terms of crime prevention. Moreover, as we have previously seen (Chap. 4), such considerations would also be relevant within a retributivist framework focussing on proportionality.

That one should use a type of sanction which *ceteris paribus* is preferable in terms of crime prevention or which may cause less harm, is theoretically a trivial point. As long as one continues to add the *ceteris paribus* clause, this is not a view that merits closer scrutiny or which would cause any philosophical

[11] As we have previously seen, one of the reasons that could be given in favour of the view that treatment by neurointerventions need not require informed consent is that this treatment may contribute in preventing significant harm to other people (future crime victims). However, if in practice such treatment is administered in cases where there is no crime preventive effect, because those treated would not have offended anyway, then this argument falls apart.

controversy. However, when we turn from penal theory to penal practice, the point becomes much less trivial. Many theorists have underlined that the resources that are being used on long prison terms could be used in much better ways. For instance, Tonry summarizes a comparison of the current use of imprisonment and possible alternatives in the following way: 'There is good evidence that imprisonment fails to reduce later offending and may increase it. There is good evidence that well-designed, well-targeted, well-resourced, and well-run treatment programs can modestly reduce later offending. Treatment programs cost much less to operate than prisons and are less likely themselves seriously to damage offenders' likelihood of living better lives later on' (Tonry 2016, p. 459). Tonry's conclusion is, contrary to what is the case, that the American criminal justice system should use prisons much more sparingly and that many offenders should be diverted into treatment programmes and community penalties (2016, p. 460). The reason for emphasizing this here is to underline that what seems ethically trivial, namely, that the criminal justice system should choose the best or most efficient alternatives, is something that cannot be taken for granted in actual penal practice.[12] But if a system is in this regard irrational, then it may of course also have implications if neurointerventions of some form were to be introduced within such a system. It would be wrong to use this sort of treatment if there were alternatives—such as programmes focussing on resocialization and reintegration into the community—which are preferable in terms of the crime preventive effect; or if there were alternatives which, everything else being equal, would be less harmful to the offender. However, as the criminal justice system actually works there are reasons to fear that such wrongful use of neurointervention might well occur. In practice, the absence of proper considerations of preferable alternatives is already part of the system.

What we have seen is that if we consider the question as to what would happen if compulsory use of neurointerventions were to be administered within a criminal system such as the one in the US, there are reasons to fear that such treatment might be wrongfully used even if *in theory* there would be some cases in which the treatment could be justified. From any moral perspective it would be wrong if such treatment, in the way outlined above, is imposed on someone who should not be treated or if there exist

[12] For considerations on the political reluctance in England and Wales to promote alternatives to incarceration, such as community orders, despite the fact that such orders have been proven to be associated with lower rates of reoffending, see e.g. Roberts and Ashworth (2016, p. 348).

alternatives that are morally preferable.[13] However, as indicated, the tendencies that could foster such wrongful use need not be dreamt up by imaginative minds, they already prevail in contemporary criminal justice practice. In fact, there is a further aspect of the irrationality of much criminal justice practice that might add to this picture, namely, the conspicuous absence of scientific evidence when new practices and legislations have been decided and enacted. If crime preventive neurointerventions were introduced in criminal justice practice without drawing heavily on various sorts of scientific evidence, then it is not difficult to imagine cases where such interventions would be administered on offenders who should not have been treated at all, where treatment would be administered despite the existence of preferable alternatives, or where such treatment would in other ways be wrongfully used. However, the lack of scientifically-informed decision-making is one of the unfortunate features of real-life criminal justice practice. In Tonry's words: 'Mandatory minimum, three-strikes, truth-in-sentencing, 'dangerous offender', 'sexual psychopath' and LWOP laws were enacted not on the basis of research findings, cost-benefit studies, impact projections, or meta-analyses, but because policy makers believed them to be intuitively plausible, morally appropriate, or politically expedient' (Tonry 2016, p. 451). Under such circumstances, it would seem reckless to offer neurointerventions as yet another option to the set of compulsory methods for dealing with crime.

(C) *Treatment as punishment*. A further aspect of the ethical discussion of the compulsory use of neurointerventions is the question of whether such interventions could be imposed as punishment. As we have seen, there are several reasons that could motivate this, at first glance, surprising discussion. What I have argued, on the ground of what is standardly presented as the necessary and sufficient conditions for something to constitute a punishment, is that there is no reason to reject the possibility that a neurointervention could function as a punishment. However, this only constituted the first part of the discussion. Penal theorists would usually hold that, even if something satisfies the definitional criteria for it to constitute a punishment, there are still further conditions that would have to be satisfied in order for a punishment to be morally justified. This led to the discussion as to whether neurointerventions could possibly satisfy any of the further requirements that penal theorists have defended. It was argued that

[13] That is, at least as long as the treatment does not constitute the less evil.

some types of neurointervention, if properly imposed, could satisfy the extra conditions. But this naturally prompts the question as to whether there are reasons to believe that these conditions would also be satisfied if we imagine that neurointerventions were to be imposed as punishment within a penal system such as the current one described in this chapter. There are several reasons to believe this would not be the case.

The first category of additional conditions that have been advocated concerns the type of punishment imposed on offenders. As we have seen, some theorists have stressed the moral significance of an attention requirement: a punishment should make it possible for the offender to focus on his misdeed by somehow bringing home to the offender the nature and implications of the crime. Another and related condition was the rationality requirement: a punishment must treat the offender as a rational being and address him by engaging with his rational capacities. Yet another requirement has been that a proper punishment should open up the possibility of a remorseful reaction in the offender. What I argued was that, even though there are of course ways of using neurointerventions as punishment that would fail to satisfy such requirements, there are also some types of neurointervention that could serve these functions. In particular, neurointerventions that improve the cognitive capacities of the offender might either in themselves or as part of some sort a punitive packages help fulfil these requirements. But are there any reasons to believe that neurointerventions would be used only in such ways if they were implemented within the framework of a penal system that is irrational in the suggested sense? It is far from clear that the answer is in the affirmative. Many of the theorists who have defended the outlined requirements have often had a critical view of penal practice, arguing that the requirements are certainly not always satisfied—in fact, they are often not satisfied—in the way the punitive systems currently work. But if that is the case, then there is no reason to believe that neurointerventions, if imposed as punishment, will suddenly be inflicted in accordance with the requirements. On the contrary, there seems to be reason to fear that this would not be the case.

This problem is even more acute if we turn to the type of requirement that does not concern types of sanction, but the severity of punishment. As we have seen, retributively-minded theorists believe that the severity of a punishment should be determined on the ground of considerations of proportionality or, at least, that proportionality places constraints on the severity of acceptable punishment. If we were to imagine different types of neurointervention implemented as punishment within a penal system, such as the one currently existing in the United States or in other Western countries, are there reasons to suggest that a proportionality requirement would be observed? The answer is clearly in the negative. As we have seen, the view of most theorists is that punishments are already being used in a disproportionate manner. Thus, there is no reason to believe that this would be any better if neurointerventions were to function as yet another type of punishment. Therefore, neither when it comes to the requirements that have been defended concerning types of punishment, nor with regard to the requirement on punitive severity, does penal practice currently give reason to believe that neurointerventions would be properly used if they were introduced among the range of punishment options.

(D) *The role of physicians.* The overall purpose of the discussion of the role of physicians, as we have seen, was to examine whether considerations concerning physician participation would in themselves constitute a ground for rejecting criminal justice use of neurointerventions. If treatment by neurointerventions presupposes involvement of physicians and if such involvement is morally unacceptable, then considerations of physician participation are sufficient to settle the whole discussion by morally blocking acceptable use of such interventions. What we have seen is that some of the arguments that have been presented in the parallel discussion of physician involvement in the death penalty could easily be dismissed. For instance, a legalist argument based merely on a reference to the Hippocratic oath turned out to be insufficient for a number of practical reasons. Moreover, there were other arguments that could be rejected because they did not deliver a general basis for the conclusion that neurointerventions would be unacceptable. This was the case with regard to objections based on the 'do no harm' *dictum* or on the particular relationship between physicians and patients. That these objections did not provide a general background for rejecting the use of neurointerventions means that there could be cases in which the use of neurointerventions for crime preventive purposes would be consistent with the moral content of these prescriptions. But if this theoretical conclusion is correct, what will follow if the objections are being rephrased within the framework of a penal system such as the current one that does not live up to ideal moral standards?

Consider first the 'do no harm' *dictum*. As we have seen, a general rejection of any kind of harm infliction would be morally absurd. There are various sorts of clinical treatment that cause some harm to patients. A more reasonable interpretation of the *dictum*, therefore, is to understand it as prohibiting treatment that causes a net surplus of harm, everything considered (i.e. when weighed against the benefits). However, as argued, in this interpretation it is no longer clear that the *dictum* would imply a rejection of all kinds of neurointervention. An intervention which causes some immediate harm, but which prevents the offender from engaging in criminal activity in the future, may benefit this person all things considered. And a neurointervention could also constitute a benefit if it were offered as an alternative to an otherwise more harmful standard punishment. However, these answers cannot be upheld if neurointerventions are used on the wrong group of offenders, that is, on individuals who would not have recidivated anyway, or if neurointerventions are administered despite the fact that there are less harmful alternatives. But as we have seen above, this is precisely what characterizes the use of many criminal sanctions in current penal practice. In numerous cases, offenders are being placed behind bars on the ground of unreliable risk assessments or despite the existence of alternative sanctions that are preferable in terms of crime preventions (and in general less harmful). Therefore, if it is plausible to maintain that physician participation should be constrained in accordance with the 'do no harm' *dictum* then, on the one hand, this does not suffice to establish that involvement in the criminal justice use of neurointerventions would *always* be wrong; however, on the other hand, given the way the current penal system works, there may well turn out to be a conflict between physician involvement in this sort of treatment and what the *dictum* prescribes.

The same is the case if we turn to one of the other standard objections that have been presented against different sorts of the non-clinical use of medical expertise; namely, the one based on the significance of the physician-patient relationship. As an objection against physician participation in the use of neurointervention, the argument could be interpreted in two ways. The first would be to contend that physicians should only be concerned with patients and that offenders do not belong in this category (as long as they are not medically impaired). As was argued, though, this objection runs counter to what is already accepted as part of the work of physicians and it is hard to explain why extra-patient involvement should be wrong if this has highly desirable consequences. The second interpretation would be to hold that offenders *should* be regarded as patients, but to contend that the patient-physician relationship forbids the sacrificing of patient interests in the attempt to realize broader societal interests. However, as was argued in relation to the 'do no harm' principle, the use of neurointerventions for crime preventive purposes need not involve a conflict between state interests and offender interests. Moreover, there may be cases where it would be acceptable to sacrifice patient interests in order to protect societal interests. But if the use of neurointervention is harmful to an offender, and if this treatment is used in cases in which there is no broader societal interest that is protected—such as if a treated offender would not have reoffended anyway—then the objection could still be upheld against such instances of physician participation. Therefore, once again, there may well be a major difference between what an objection implies in a general analysis of the use of neurointerventions and what it implies in relation to actual use under non-ideal circumstances.

But then what should we say with regard to physician involvement in the administration of neurointervention if such treatment were introduced within a criminal justice system that is irrational in the sense described above? Since the issue hinges very much on uncertain empirical questions, a firm answer is not warranted. However, two things should be noted. A reason against this sort of physician participation might be that it could result in an erosion of trust in physicians. Though this is an objection that is hard to back up empirically, there is also no ground for dismissing it. Furthermore, and more generally, what I suggested was that the rightness or wrongness of physician involvement should be assessed on the ground of a bifurcate analysis. Following what I called the 'dependency view', one should ask, first, whether the activity in question is morally desirable and, subsequently, whether and how this activity would be carried out if physicians were to abstain from undertaking the job. But this implies that, if some of the mechanisms that have led to unacceptable over-punishment would also lead to over-treatment in the sense that treatment is imposed on someone who should not have been treated or is imposed despite the existence of preferable alternatives then, as suggested, it may be the case that the use of neurointerventions is morally undesirable. The next question thus is, whether this treatment would be carried out without the involvement of physicians. The answer, as we have seen—in contrast to the death penalty which can be performed without physician participations—is that this is unlikely to be the case.[14] The administration of neurointerventions would require some sort of physician involvement. But

[14] There has been some discussion as to whether it is possible to administer lethal injections without the involvement of physicians. However, traditional ways of imposing the death penalty (e.g. by shooting or hanging) do not presuppose medical expertise.

in that case, the dependency view implies that the reasons against this activity would also constitute reasons in relation to the question of physician participation and, hence, that it would be wrong for physicians to engage in this treatment. Thus, though we should be cautious in our drawing of conclusions, it seems that under non-ideal circumstances it is not unlikely that the use of neurointervention on offenders would be morally blocked by the fact that physicians should stay out of this sort of treatment.

In summary, what the above considerations show is that a number of the objections to the use of neurointerventions for crime preventive purposes, which I have previously argued were invalid because they did not comprise all types of neurointervention or all ways of using such interventions, may have more force when they are presented within the framework of a penal system such as the one initially described. Moreover, I have suggested that some of the features characterizing current penal practice may lead to a wrongful use of neurointerventions. These characteristics may imply the administration of treatment to offenders who should not be treated (but who will suffer the side-effects); the use of treatment despite the fact that there exist preferable alternatives; and, more generally, the implementation of treatment without the possession of sufficient scientific evidence. As mentioned, one cannot know for sure that these features of the use of punishment will spill over into the way neurointerventions would be applied. However, as indicated, there are some mechanisms that make it reasonable to fear that this would be the case. Moreover, there is a further way in which this judgment could be substantiated. Rather than engaging in hypothetical considerations of what would follow if the characteristics of the use of punishment are extrapolated to the administration of neurointerventions, one could consider whether any of the suggested trends can be detected in the way some neurointerventions are already in use. As we have seen earlier, there are today a number of US states—as well as jurisdictions in other countries—that have passed legislation authorizing the use of chemical castration of sex offenders. Can any of the suggested traits be found in the ways in which this treatment is being administered? Though I cannot here engage in a thorough analysis of the current use of chemical castration—which would be complicated by the fact that the legislation differs between the states that have authorized it—there are indeed some reasons indicating that the answer is in the affirmative.[15] Here are a few examples.

[15] The first state to authorize the use of chemical castration was California in 1996. For a comparison of the legislation that has subsequently been passed in other US states, see Scott and Holmberg (2003).

There are reasons to fear that chemical castration is administered on those who should not have been thus treated because they do not possess a risk. First, though there are some sex offenders who are excessively dangerous—such as those who are both antisocial and paraphilic—the majority of sex offenders covered by castration statutes have no sexual disorders, and some meta-studies have indicated the risk of recidivism in this group to be low (see Kernsmith et al. 2016; Stinneford 2006). Furthermore, though there are differences between the states, a treatment that has been initiated is likely to continue, in some cases, for the life of the offender. Stinneford notes that 'it is difficult to imagine a scenario in which the state, after deciding that an offender needed chemical castration, would later decide that it was 'no longer necessary'. Therefore, in most cases, the offender will be subjected to a life term of chemical castration' (2006, p. 580). To what extent this is justified where offenders not suffer from sexual disorders is an open question. Second, in considering the alternatives to chemical treatment, Stinneford further notes that 'several studies indicated that cognitive behavioral theory is as effective as chemical castration in preventing recidivism' (2006, p. 575). More recent studies have indicated that chemical castration has a greater effect than alternative types of treatment (Kim et al. 2016). But it is interesting that these results have been reached many years after treatment by chemical castration has been introduced. Third, there are several ways in which the administration of chemical castration has been held, by commenters, not to be scientifically informed. For instance, in some states there is no statutory requirement that a physician determines a treatment to be medically appropriate and safe (Stinneford 2006, p. 578). And in those states which require the involvement of healthcare professionals, medical reviews occur close to the time of sentencing. However, since treatment is often not initiated until briefly before the time of release from prison, there may in practice be a major gap—in some cases decades—between a medical review and the treatment. As Stinneford remarks: 'under such circumstances, even a mere determination that the defendant can physically tolerate the medication will no longer be valid at the time treatment commences' (2006, p. 583). More generally, Rice and Harris have summarized their review of the scientific literature on chemical castration in the following way:

> Little is known about its effects on sexual or violent recidivism among sex offenders who do not freely request it. Moreover, little is known about the characteristics of those who volunteer for (rather than refuse) ADT

> [androgen deprivation therapy], especially such risk-relevant characteristics as psychopathy or high actuarial risk scores. Little is known about the long term effects of ADT on sexual behavior in general and sexual recidivism in particular. Little is known about the long term health effects of ADT. (Rice and Harris 2011, p. 328)[16]

The authors consequently contend that much more research is needed before 'ADT has a sufficient scientific basis to be relied upon as a principal component of sex offender treatment' (2011, p. 328). It is remarkable that this conclusion is drawn about 15 years after California, as the first state, authorized the use of chemical castration for sex offenders. Thus, even though much more can be said on the current legislation and practice regarding treatment of sex offenders, there is—as these few points indicate—no ground for contending that the concerns that have been presented in this chapter should be put to rest in the light of the current use of neurointerventions.[17]

A final point merits a comment. As is now clear, the considerations in this chapter rest on a bleak view of current penal practice and the way in which criminal sanctions are being administered to offenders. It is this judgement that forms the background to the pessimism with regard to what could be expected if treatment in the form of neurointerventions were to be put into practice. However, it could be held that pessimistic judgements concerning current penal practice might lend some support for the opposite conclusion. After all, if current penal practice is, in various ways, on the wrong track, and has been so for many years, then it could be the case that new ways of dealing with offenders in the form of new treatment options might constitute *the least evil*. For instance, even if the administration of a neurointervention would in several regards be morally undesirable, the treatment could still be preferable if it implies that fewer offenders would be severely over-punished. However, though I believe that there is much to be said in favour of thinking in terms of the least evil when one

[16] It is important to note that, even though the authors hold that little is known about the long-term health effects, there is a host of established harmful physical and mental health risks of chemical castration including, for example, an increased risk of fractures and diabetes mellitus, an increased risk in cardiovascular morbidity, and in symptoms of depression (see Rice and Harris 2011).

[17] Commenting on the current castration laws in the United States, Rice and Harris contend that 'the major impetus may have been political—sex offenders, especially pedophiles, are reviled by much of the lay public, and politicians who push for such laws clearly gain political favor by doing so' (Rice and Harris 2011, p. 326).

is assessing alternatives under non-ideal circumstances, I do not know if there are any reasons supporting such a forecast as to what would happen if neurointerventions were to a greater extent to become part of criminal justice practice. Thus, though it is important to keep in mind that under non-ideal circumstances a treatment scheme should not only be assessed in terms of whether it is ideally desirable, but sometimes also in terms of whether it may contribute to turning an undesirable state of affairs into one that is less undesirable, there is not sufficient ground to feel confident that this would be the result if such treatment were to be implemented. In short, the perspective is important, but in the absence of any kind of empirical support it does little to mitigate the force of the concerns generated by the irrationality of current penal practice.

Conclusion

Should treatment by neurointerventions be used as an instrument to prevent offenders engaging in future criminal activity? This question, which has constituted the pivotal point in the discussion throughout this book, is ambiguous in a way that is not always clearly realized in the neuroethical discussion that has evolved in recent years.

On the one side, the question can be interpreted as asking whether it would in principle or ever be acceptable to administer this type of treatment on offenders. If the side-effects are assumed not to be too adverse and if such treatment would reduce an offender's propensity for crime, would it then be acceptable (or morally required) to administer this treatment or would there still be reasons to regard the treatment as morally objectionable? This is the question that philosophers have typically posed. The answer obviously presupposes further clarification. Is the treatment freely accepted or administered in a compulsory manner? And what type of neurointervention are we considering? However, as has been argued in the previous chapters, many of the principled objections that have been raised against crime preventive use of neurointervention are premature. More precisely, I have suggested that they are either premised on morally implausible principles—for instance, views that are inconsistent with what is regarded as an acceptable (or required) use of criminal sanctions—or they are *non sequitur*, in the sense that they are not inconsistent with the use of some types of neurointervention or certain ways of administering such interventions.

On the other side, the question is also susceptible to a different interpretation, namely, whether it would be morally desirable to administer treatment by neurointervention within criminal justice systems as they currently are (or will be in the future). It is this question that has constituted the main focus of the present chapter and which, apparently, is closer to how law scholars and social scientists tend to approach the issue. As we have seen, the answer obviously depends upon what characterizes the context within which the treatment is implemented. Such considerations are conditional on many empirical assumptions and, as underlined, will be highly speculative. This, of course, means that one should be careful not to overstate the strength of the conclusions. However, as has been argued, there are reasons to fear that it will in various ways not be morally desirable to implement such treatment within a context such as the current criminal justice systems in the United States and other Western countries.

Considerations of both interpretations of the initial question are clearly important. Principled discussions of whether it could ever be justified to use neurointerventions in the attempt to prevent crime will help to clarify the main ethical reasons for and against the application of this technology and, thereby, also to specify the conditions that have to be satisfied in order for the treatment to be acceptable. Considerations of what could be expected if the treatment were to be put into practice are obviously of equal importance, since the goal of addressing fundamental ethical questions is not per se some sort of philosophical puzzle solving but, ultimately, to devise solutions to real-world challenges. If there seems to have been a move from a more optimistic and advocatory approach to the use of neurointerventions in the first and main part of the chapters to a much more pessimistic approach in the final chapters, then this basically reflects the major distance between in-principle considerations of how things ought to be and a view of the criminal justice system as an institution which in practice deviates significantly from what is ideally desirable.

Bibliography

Aarons, L. (1972). 'Brain Surgery Is Tested on Three California Convicts'. *Washington Post*, 25 February, p. 1.

Agar, N. (2014). 'A Question About Defining Moral Bioenhancement'. *Journal of Medical Ethics* 40: 369–270.

Agar, N. (2015). 'Moral Enhancement is Dangerous'. *Journal of Medical Ethics* 41: 343–345.

Anderson, S. (2011). 'Coercion'. In E. Zelta (ed.), *Stanford Encyclopaedia of Philosophy* (online: https://plato.stanford.edu/).

Andreasson, E. and I. Skärsäter (2012). 'Patients Treated for Psychosis and Their Perceptions of Care in Compulsory Treatment: Basis for an Action Plan'. *Journal of Psychiatric and Mental Health Nursing* 19: 15–22.

Annas, G. J. and L. H. Glantz (1974). 'Psychosurgery: The Law's Response'. *Boston University Law Review* 54: 249–267.

Arboleda-Florez, J. (2005). 'The Ethics of Biomedical Research on Prisoners'. *Current Opinion in Psychiatry* 18 (5): 514–517.

Baertschi, B. (2014). 'Neuromodulation in the Service of Moral Enhancement'. *Brain Topography* 27: 63–71.

Barn, G. (2016). 'Can Medical Interventions Serve as "Criminal Rehabilitation"?'. *Neuroethics* (online first).

Bateson, M. et al. (2006). 'Cues of Being Watched Enhance Cooperation in a Real-World Setting'. *Biology Letters* 2: 412–414.

Baum. K. (2001). 'To Comfort Always: Physician Participation in Executions'. *N.Y.U. Journal of Legislation and Public Policy* 5: 47–81.

Baumgartner et al. (2008). 'Oxytocin Shapes the Neural Circuitry of Trust and Trust Adaption in Humans'. *Neuron* 58 (4): 639–650.

Baylis, F. (2013). 'I am Who I Am: On the Perceived Threats to Personal Identity From Deep Brain Stimulation'. *Neuroethics* 6: 513–526.

Beck, B. (2015). 'Conceptual and Practical Problems of Moral Enhancement'. *Bioethics* 29: 233–240.

Beehler, R. (1982). 'Containing Violence'. *Ethics* 92 (4): 647–660.

Benn, S. I. (1967). 'Punishment'. In J. G. Murphy (ed.), *Punishment and Rehabilitation*, pp. 8–19. Belmont: Wadsworth.

Black, L. and H. Fairbrother (2008). 'The Ethics of the Elephants: Why Physician Participation in Executions Remains Unethical'. *American Journal of Bioethics* 8: 59–60.

Black, L. and R. M. Sade (2007). 'Lethal Injections and Physicians: State Law vs. Medical Ethics'. *JAMA Journal of the American Medical Association* 298: 2779–2781.

Blumenthal-Barby, J. S. (2014). 'A Framework for Assessing the Moral Status of Manipulation'. In M. C. C. Weber (ed.), *Manipulation*, pp. 121–134. Oxford: Oxford University Press.

Boire, R. G. (2010). 'State-Imposed Brain Intervention: The Case for Pharmacotherapy for Drug Abuse'. In M. Farah (ed.), *Neuorethics: An Introduction with Readings*, pp. 281–300. Cambridge: MIT Press.

Bolt, L. L. E. (2007). 'True to Oneself? Broad and Narrow Ideas on Authenticity in the Enhancement Debate'. *Theoretical Medicine and Bioethics* 28: 285–300.

Bomann-Larsen, L. (2013). 'Voluntary Rehabilitation? On Neurotechnological Behavioural Treatment, Valid Consent and (In)appropriate Offers'. *Neuroethics* 6: 65–77.

Boonin, D. (2008). *The Problem of Punishment*. New York: Cambridge University Press.

Brant, J. (1974). 'Behavior Modification as a Potential Infringement of Prisoners' Right to Privacy'. *New England Journal on Prison Law* 1: 180–201.

Breggin, P. (1973). 'The Second Wave of Psychosurgery'. *Mental Health* 57: 10–13.

Breggin, P. (1975). 'Psychosurgery for Political Purposes'. *Duquesne Law Review* 13: 841–862.

Brenninkmeijer, J. (2010). 'Taking Care of One's Brain: How Manipulating the Brain Changes People's Selves'. *History of the Human Sciences* 23: 107–126.

Brody, H. and M. Wardlaw (2008). 'Two Gorillas in the Death Penalty Room'. *American Journal of Bioethics* 8: 53–54.

Brooks, T. (2012). 'Moral Frankensteins'. *AJOB Neuroethics* 3: 28–30.

Browne, J. (2007). 'Rooted in Slavery: Prison Labor Exploitation'. *Race, Poverty, and the Environment* 14 (1): 42–44.

Bruce, C. R. (2008). 'Lethal Injections: Legal Extensions and Implications of Do No Harm'. *American Journal of Bioethics* 8: 58–59.

Bublitz, J. C. (2011). 'If Man's True Palace Is His Mind, What Is Its Adequate Protection? On a Right to Mental Self-Determination and Limits of Interventions into Other Minds'. In B. van den Berg and L. Klaming (eds.), *Technologies on the Stand. Legal and Ethical Questions in Neuroscience and Robotics*, pp. 89–114. Nijmegen: Wolf legal Publishers.

Bublitz, J. C. (2016). 'Moral Enhancement and Mental Freedom'. *Journal of Applied Philosophy* 33 (1): 88–106.

Bublitz, J. C. (2018). '"The Soul is the Prison of the Body"—Mandatory Moral Enhancement, Punishment and Rights Against Neuro-Rehabilitation'. In D. Birks and T. Douglas (eds.), *Treatment for Crime: Philosophical Essays on Neurointerventions in Criminal Justice*, pp. 289–320. Oxford: Oxford University Press.

Bublitz, J. C. and R. Merkel (2009). 'Autonomy and Authenticity of Enhanced Personality Traits'. *Bioethics* 23: 360–374.

Bublitz, J. C. and R. Merkel (2014). 'Crimes Against Minds: On Mental Manipulations, Harms and a Human Right to Mental Self-Determination'. *Criminal Law and Philosophy* 8: 51–77.

Buller, T. (2013). 'Neurotechnology, Invasiveness and the Extended Mind'. *Neuroethics* 6: 593–605.

Burdon, W. M. and C. A. Gallagher (2002). 'Coercion and Sex Offenders'. *Criminal Justice and Behavior* 29: 87–109.

Burgess, A. (1972). *A Clockwork Orange*. Harmondsworth: Penguin Books.

Butler, T. et al. (2010). 'Reducing Impulsivity in Repeat Violent Offenders: An Open Trial of a Selective Serotonin Reuptake Inhibitor'. *Australian and New Zealand Journal of Psychiatry* 44: 1137–1143.

Bylinsky, G. (1973). 'New Clues to the Causes of Violence'. *Fortune*, Jan: 134–136.

Cabrera, L. Y. et al. (2014). 'Ethics of the Electrified Mind'. *Brain Topography* 27 (1): 33–45.

Caplan, A. L. (2006). 'Ethical Issues Surrounding Forced, Mandated, or Coerced Treatment'. *Journal of Substance Abuse* 31 (2): 117–120.

Caplan, A. L. (2007). 'Should Physicians Participate in Capital Punishment?'. *Mayo Clinic Proceedings* 82: 1047–1048.

Caplan, A. L. (2008). 'Editorial'. *Addiction* 103: 1919–1921.

Casey, B. P. (2015). 'The Surgical Elimination of Violence? Conflicting Attitudes Towards Technology and Science During the Psychosurgery Controversy of the 1970s'. *Science in Context* 28 (1): 99–129.

Center for Cognitive Liberty and Ethics (2004). 'Threats to Cognitive Liberty: Pharmacotherapy and the Future of Drug War'. Available online at: http://m-bikeshop.com/library/download/asin=0:00265993&type=stream

Chorover, S. L. (1974). 'Psychosurgery: A Neuropsychological Perspective'. *Boston University Law Review* 54: 231–248.

Choy, O. et al. (2018a). 'Stimulation of the Prefrontal Cortex Reduces Intentions to Commit Aggression: A Randomized, Double-Blind, Placebo-Controlled, Stratified Parallel-Group Trial'. *The Journal of Neuroscience* (online first).

Choy, O. et al. (2018b). 'Benign Biological Interventions to Reducer Offending'. *Neuroethics* 2018 (online first).

Clark, P. A. (2006). 'Physician Participation in Executions: Care Giver or Executioner?'. *Journal of Law, Medicine and Ethics* 34: 95–104.

Clarkson, P. A. (1939). 'Have You A Wrong Way Brain?'. *Popular Science* 135: 45–48.

Craig, J. N. (2016). 'Incarceration, Direct Brain Interventions, and the Right to Mental Integrity—a Reply to Thomas Douglas'. *Neuroethics* 9 (2): 107–118.

'Crime Cure?'. (1947). *Time*, July 14.

Crockett, M. L. et al. (2010). 'Serotonin selectively Influences Moral Judgment and Behaviour Through Effects on Harm Aversion'. *Proceedings of the National Academy of Science of the United States of America* 107: 17433–17438.

Crockett, M. J. (2014). 'Moral Bioenhancement: A Neuroscientific Perspective'. *Journal of Medical Ethics* 40 (6): 370–371.

Croxson, P. (2012). 'Acetylcholine Facilitates Recovery of Episodic Memory after Brain Damage'. *The Journal of Neuroscience* 32 (40): 13787–13795.

Cunha-Bang, S. et al. (2016). 'Serotonin 1B Receptor Binding Is Associated With Trait Anger and Level of Psychopathy in Violent Offenders'. *Biological Psychiatry* 82 (4): 1–8.

Dagger, R. (2012).'Playing Fair with Recidivists'. In C. Tamburrini and J. Ryberg (eds.), *Recidivist Punishments*, pp. 41–62. Lanham: Lexington Books.

Davis, M. (1992). *To Make Punishment Fit the Crime*. United States of America: Westview Press.

Davis, M. (2012). 'Recidivist Penalties Revisited'. In C. Tamburrini and J. Ryberg (eds.), *Recidivist Punishments*, pp. 23–40. Lanham: Lexington Books.

Day, A. et al. (2004). 'Coerced Offender Rehabilitation—A Defensible Practice?'. *Psychology*, Crime *and Law* 10 (3): 259–269.

Degrazia, D. (2000). 'Prozac, Enhancement, and Self-Creation'. *Hastings Center Report*, March–April 2000: 34–40.

Degrazia, D. (2005). 'Enhancement Technologies and Human Identity'. *Journal of Medicine and Philosophy* 30 (3): 261–283.

Degrazia, D. (2012). 'Moral Enhancement, Freedom, and What We (Should) Value in Moral Behaviour'. *Journal of Medical Ethics* 40: 361–368.

Dekkers, W et al. (2005). 'Bodily Integrity and Male and Female Circumcision'. *Medicine, Health Care and Philosophy* 8: 179–191.

Douglas, T. (2008). 'Moral Enhancement'. *Journal of Applied Philosophy* 25 (3): 228–245.

Douglas, T. (2014). 'Moral Bioenhancement, Freedom and Reasoning'. *Journal of Medical Ethics* 40: 359–360.

Douglas, T. et al. (2013). 'Coercion, Incarceration, and Chemical Castration: An Argument from Autonomy'. *Bioethical Inquiry* 10: 393–405.

Douglas, T. (2019). 'Nonconsensual Neurocorrectives and Bodily Integrity: A Reply to Shaw and Barn'. *Neuroethics* 12 (1): 107–119.

Duff, R. A. (1986). *Trials and Punishment*. Cambridge: Cambridge University Press.

Duff, R. A. (1998). 'Dangerousness and Citizenship'. In A. Ashworth and M. Wasik (eds.), *Fundamentals of Sentencing Theory*, pp. 141–164. Oxford: Clarendon Press.

Duff, R. A. (2001). *Punishment, Communication, and Community*. Oxford: Oxford University Press.

Duff, R. A. (2005). 'Punishment, Dignity, and Degradation'. *Oxford Journal of Legal Studies* 25: 141–155.

Duff, R. A. and D. Garland (1998). 'Introduction'. In A. Duff and D. Garland (eds.), *Punishment*, pp. 1–42. New York: Oxford University Press.

Duff, R. A. and Z. Hoskins (2017). 'Legal Punishment'. In A. N. Zalta (ed.), *Stanford Encyclopedia of Philosophy* (online: https://plato.stanford.edu/).

Dworkin, G. (2002). 'Patients and Prisoners: The Ethics of Lethal Injection'. *Analysis* 62: 181–189.

Earp, B. D. et al. (2017). 'Moral Neuroenhancement'. In S. Johnson and K. Rommelfanger (eds.), *Routledge Handbook of Neuroethics*, pp. 166–184. New York: Routledge.

Elliott, C. (1999). *A Philosophical Disease: Bioethics, Culture and Identity*. New York: Routledge.

Enggist, S. et al. (2014). *Prison and Health*. WHO publication.

Erler, A. (2011). 'Does Memory Modification Threaten Out Authenticity?'. *Neuroethics* 4: 235–249.

Eskine, K. J. et al. (2011). 'A Bad Taste in the Mouth'. *Psychological Science* 22 (3): 295–299.

Farah, M. (2007). 'Emerging Ethical Issues in Neuroscience'. In W. Glannon (ed.), *Defining Right and Wrong in Brain Science*, pp. 37–52. New York: Dana Press.

Farah, M. (ed.) (2010). *Neuorethics: An Introduction with Readings*. Cambridge: MIT Press.

Faria, M. A. (2013). 'Violence, Mental Illness, and the Brain—A Brief History of Psychosurgery: Part 3—From Deep Brain Stimulation to Amygdalotomy for Violent Behavior, Seizures, and Pathological Aggression in Humans'. *Surgical Neurology International* 4: 1–27.

Farrington, D. P. (2002). 'Key Results from the First 40 Years of the Cambridge Study in Delinquent Development'. In T. P. Thornberry and M. Krohn (eds.), *Taking Stock of Delinquency*, pp. 137–184. New York: Kluwer/Plenum.

Fazel, S. et al. (2012). 'Use of Risk Assessment Instruments to Predict violence and Antisocial Behaviour in 73 Samples Involving 24827 People: Systematic Review and Meta-analysis'. *British Medical Journal* 345: 1–12.

Feinberg, J. (1965). 'The Expressive Function of Punishment'. The Monist 49. Reprinted in J. Feinberg (1970) *Doing and Deserving*. Princeton: Princeton University Press.

Feinberg, J. (1983). 'Noncoercive Exploitation'. In R. Sartorious (ed.), *Paternalism*, pp. 201–235. Minneapolis: University of Minnesota Press.

Feinberg, J. (2003). *The Moral Limits of the Criminal Law Volume 3: Harm to Self*. Oxford: Oxford Scholarship Online.

Feldman, M. P. (1966). 'Aversion Therapy for Sexual Deviations: A Critical Review'. *Psychological Bulletin* 65 (2): 65–79.

Fletcher, G. (1978). *Rethinking Criminal Law*. Boston: Little, Brown and Company.
Focquaert, F. (2014). 'Mandatory Neurotechnological Treatment: Ethical Issues'. *Theoretical Medicine and Bioethics* 35: 59–72.
Forsberg, L. and T. Douglas (2017). 'Anti-Libidinal Interventions in Sex Offenders: Medical or Correctional?'. *Medical Law Review* 24 (4): 453–473.
Franzini, A. et al. (2013). 'Deep-rain Stimulation for Aggression and Disruptive Behavior'. *World Neurosurgery* 80: 11–14.
Garland D. (2001). 'Introduction: The Meaning of Mass Imprisonment'. *Punishment and Society* 3 (1): 5–7.
Gawande, A. (2006). 'When Law and Ethics Collide—Why Physicians Participate in Executions'. *The New England Journal of Medicine* 354: 1221–1229.
Gaylin, W. and H. Blatte (1975). 'Behavior Modification in Prisons, *American Criminal Law Review* 13: 11–35.
Gesch, C. B. et al. (2002). 'Influence of Supplementary Vitamins, Minerals and Essential Fatty Acids on the Antisocial Behaviour of Young Adult Prisoners. Randomized, Placebo-Controlled Trial'. *The British Journal of Psychiatry* 181: 22–28.
Gilbert, F (2015a). 'A Threat to Autonomy? The Intrusion of Predictive Brain Implants'. *AJOB Neuroscience* 6 (4): 4–11.
Gilbert, F. (2015b). 'Self-Estrangement and Deep Brain Stimulation: Ethical Issues Related to Forced Explanation'. *Neuroethics* 8: 107–114.
Gilbert, F. et al. (2017). 'I Miss Being Me: Phenomenological Effects of Deep Brain Stimulation'. *AJOB Neuroethics* 8 (2): 96–109.
Ginneken, E. (2019). 'The Use of Risk Assessment in Sentencing'. In Keijser et al. (eds.), *Predictive Sentencing*, pp. 9–32. Oxford: Hart Publishing.
Ginsberg, Y. and N. Lindefors (2012). 'Methylphenidate Treatment of Adult Male Prison Inmates with Attention-Deficit Hyperactivity Disorder: Randomised Double-blind Placebo-Controlled Trial with Open-Label Extension'. *British Journal of Psychiatry* 200: 68–73.
Giubilini, A. and S. Sanyal (2015). 'The Ethics of Human Enhancement'. *Philosophy Compass* 10: 233–243.
Glannon, W. (ed.). *Defining Right and Wrong in Brain Science*. New York: Dana Press.
Glannon, W. (2008). 'Psychopharmacological Enhancement'. *Neuroethics* 1: 45–54.
Glannon, W. (2014). 'Neuromodulation, Agency and Autonomy'. *Brain Topography* 27: 46–54.
Glenn, A. L. and A. Raine, 'Neurocriminology: Implications for the Punishment, Prediction and Prevention of Criminal Behaviour'. *Nature Reviews Neuroscience* 15: 54–63.
Goodin, R. (1987). 'Exploiting a Situation and Exploiting a Person'. In A. Reeve (ed.), *Modern Theories of Exploitation*, pp. 166–200. London: Sage.
Greely, H. T. (2008). 'Neuroscience and Criminal Justice: Not Responsibility but Treatment'. *Kansas Law Review* 56: 1103–1138.
Greely, H. T. (2012). 'Direct Brain Interventions to "Treat" Disfavoured Human Behaviors: Ethical and Social Issues'. *Clinical Pharmacology and Therapeutics* 91: 163–165.
Green, W. (1986). 'Depo-Provera, Castration, and the Probation of Rape Offenders: Statutory and Constitutional Issues'. *University of Dayton Law Review* 12 (1): 1–26.
Groner, J. I. (2002). 'Lethal Injection: A Stain on the Face of Medicine'. *British Medical Journal* 325: 1026–1029.

Groner, J. I. (2008). 'The Hippocratic Paradox: The Role of the Medical Profession in Capital Punishment in the United States'. *Fordham Urban Law Journal* 35: 883–917.

Grön, G. et al. (2005). 'Cholinergic Enhancement of Episodic Memory in Healthy Young Adults'. Psychopharmacology 182: 170–179.

Guidry, O. F. (2006). 'Message from the President: Observations Regarding Lethal Injection', June 30, available at: http://www.yesmagazine.org/issues/go-local/1561.

Harat, M et al. (2015). 'Deep Brain Stimulation in Pathological Aggression'. *Stereotactic and Functional Neurosurgery* 93: 310–315.

Harris, J. (2011). 'Moral Enhancement and Freedom'. *Bioethics* 25: 102–111.

Harris, J. (2013a). 'Moral Progress and Moral Enhancement'. *Bioethics* 27: 285–290.

Harris, J. (2013b). 'Ethics Is for Bad Guys! Putting the Moral into Moral Enhancement'. *Bioethics* 27: 169–173.

Harris, J. (2013c). 'What It Is to Be Good'. *European Review* 21: 114–122.

Harris, J. (2014a). 'Taking Liberties with Free Fall'. *Journal of Medical Ethics* 40: 371–374.

Harris, J. (2014b). '. . . How Narrow the Strait!'. *Cambridge Quarterly of Healthcare Ethics* 23: 247–260.

Hauskeller, M. (2016). 'The Art of Misunderstanding Critics'. *Cambridge Quarterly of Healthcare Ethics* 25: 153–161.

Heldman, V. C. (1974). 'Behavior Modification and Other Legal Imbroglios of Human Experimentation, *Journal of Urban Law* 52: 155–175.

Hill, T. E. (1992). *Dignity and Practical Reason in Kant's Moral Philosophy*. Ithaca: Cornell University Press.

Hollin, S and C. Bilby (2007). 'Addressing Offender Behaviour: What Works and Beyond'. In Y. Jewkeys (ed.), *Handbook of Prison*, pp. 608–629. Devon: Willan.

Holtug, N. (2009). 'Equality and the Treatment-Enhancement Distinction'. *Bioethics* 25 (3): 137–144.

Hornblum, A. E. (1997). 'They Were Cheap and Available: Prisoners as Research Subjects in the Twentieth Century America'. *British Medical Journal* 315: 1437–1441.

Horstkötter, D et al. (2012). 'Moral Enhancement for Antisocial Behavior? An Uneasy Relationship'. *AJOB Neuroethics* 3: 26–28.

Hübner, D. and L. White (2016). 'Neurosurgery for Psychopaths'. *AJOB Neuroscience* 7: 140–149.

Husak, D. (2008). *Overcriminalization: The Limits of the Criminal Law*. New York: Oxford University Press.

Husak, D. (2011). 'Lifting the Cloak: Preventive Detention as Punishment'. *San Diego Law Review* 48: 1173–1204.

Husak, D. (2019a). 'Why Legal Philosophers (Including Retributivists) Should Be Less Resistant to Risk-Based Sentencing'. In J. de Keijser et al., *Predictive Sentencing*, pp. 33–50. Oxford: Hart Publishing.

Husak, D. (2019b). 'Kinds of Punishment'. In H. M. Hurd (ed.), *Moral Puzzles and Legal Perplexities*, pp. 23–38. Cambridge: Cambridge University Press.

Illes, J. and B. Sahakian (eds.) 2011. *The Oxford Handbook of Neuroethics*. Oxford: Oxford University Press.

James, B. (1962). 'Case of Homosexuality Treated by Aversion Therapy'. *British Medical Journal* 1: 768–770.

Johansson, V. et al. (2014). 'Thinking Ahead on Deep Brain Stimulation: An Analysis of the Ethical Implications for a Developing Technology'. *AJOB Neuroethics* 5 (1): 24–33.

Johnson, J. (2009). 'A Dark History: Memories of Lobotomy in the New Era of Psychosurgery'. *Medicine Studies* 1: 367–378.
Jotterand, F. and J. Giordano (2011). 'Transcranial Magnetic Stimulation, Deep Brain Stimulation and Personal Identity: Ethics Questions and Neuroethical Approaches for Medical Practice'. *International Review of Psychiatry* 23 (5): 476–485.
Kagan, S. (1998). *Normative Ethics*. Colorado: Westview Press.
Kahneman, D. (2011). *Thinking, Fast and Slow*. New York: Farrar, Straus and Giroux.
Kant, I. (1887). *The Philosophy of Law*. Edingburgh: T. and T. Clark.
Kant, I. (1965). *The Metaphysical Elements of Justice*. J. Ladd (ed.). Indianapolis: Bobbs-Merill.
Kant, I. (2012). *Groundwork of the Metaphysics of Morals*. New York: Cambridge University Press.
Keane, M. (2008). 'The Ethical Elephant in the Death Penalty Room'. *American Journal of Bioethics* 8: 45–50.
Keijser, J., J. V. Roberts, and J. Ryberg. (2019). *Predictive Sentencing*. Oxford: Hart Publishing.
Kernsmith, P. et al. (2016). 'Fear and Misinformation as Predictors of Support for Sex Offender Management Policies'. *Journal of Sociology and Social Welfare* 39: 39–66.
Kim, B et al. (2016). 'Sex Offender Recidivism Revisited: Review of Recent Meta-analyses on the Effects of Sex Offender Treatment'. *Trauma, Violence, and Abuse* 17 (1): 105–117.
Kleinig, J. (1973). *Punishment and Desert*. The Hague: Martinus Nijhoff.
Kleinig, J. (1998). 'The Hardness of Hard Treatment'. In A. Ashworth and M. Wasik (eds.), *Fundamentals of Sentencing Theory*, pp. 273–290. Oxford: Clarendon Press.
Kolber, A. (2009). 'The Subjective Experience of Punishment'. *Columbia Law Review* 109: 182–211.
Kosfeld, M. et al. (2005). 'Oxytocin Increases Trust in Humans'. *Nature* 435 (2): 673–676.
Kraemer, F. (2013). 'Me, Myself and My Brain Implant: Deep Brain Stimulation Raises Questions of Personal Authenticity and Alienation'. *Neuroethics* 6: 483–497.
Kringelbach, M. L. and T. Aziz (2009). 'Deep Brain Stimulation'. *JAMA* 301 (16): 1705–1708.
Kuhse, H. (2000). 'Is There a Tension Between Autonomy and Dignity'. In P. Kemp et al, *Bioethics and Biolaw*, pp. 61–74. Copenhagen: Rhodos.
Lacey, N. (2007). 'Criminal Justice and Democratic Systems: Inclusionary and Exclusionary Dynamics in the Institutional Structure of Late Modern Societies'. *Center for European Studies Working Paper Series #148*.
Lacey, N. (2008). *The Prisoners' Dilemma*. Cambridge: Cambridge University Press.
Lapidus, K. A. B: et al. (2013). 'History of Psychosurgery: A Psychiatrist's Perspective'. *World Neurosurgery* 80 (3): 1–16.
Lappi-Seppälä, T. (2019). 'The Principle of Proportionality and other Values in Nordic Sentencing'. In M. Tonry (ed.), *Proportionality, Punishment, and Justice* (forthcoming).
Lawson, A. (2005). 'Disability, Degradation, and Dignity: The Role of Article 3 of the European Convention of Human Rights'. *Northern Ireland Legal Quarterly* 56: 462–491.
Lerman, D. N. (2006). 'Second Opinion: Inconsistent Deference to Medical Ethics in Death Penalty Jurisprudence'. *The Georgetown Law Journal* 95: 1941–1978.
Levy. N. 2007. *Neuroethics: Challenges for the 21st Century*. Cambridge: Cambridge University Press.
Levy, N. (2011). 'Enhancing Authenticity'. *Journal of Applied Philosophy* 28: 308–318.

Levy, N. et al. (2014). 'Are You Morally Modified? The Moral Effects of Widely Used Pharmaceuticals'. *Philosophy, Psychiatry, Psychology* 21 (2): 111–125.
Lichtenstein, P. et al. (2012). *The New England Journal of Medicine* 367: 2006–2014.
Liebling, A. and S. Maruna (2005). *The Effects of Imprisonment*. Devon: Willan.
Lippert-Rasmussen, K. (2008). 'Against Self-Ownership: There are No Fact-Insensitive Ownership Rights over One's Body'. *Philosophy and Public Affairs* 36 (1): 86–118.
Litton, P. (2013). 'Physician Participation in Executions, the Morality of Capital Punishment, and the Practical Implication of Their Relationship'. *Journal of Law, Medicine, and Ethics* 41: 333–352.
Lunze, K. et al. (2016). 'Mandatory Addiction Treatment for People Who Use Drugs: Global Health and Human Rights Analysis'. *British Medical Journal* (online first).
Maguire, E. A. et al. (2006). 'London Taxi Drivers and Bus Drivers: A Structural MRI and Neuropsychological Analysis'. *Hippocampus* 16 (12): 1091–1101.
Manenti, R. et al. (2013). 'Enhancing Verbal Episodic Memory in Older and Yong Subjects after Non-invasive Brain Stimulation'. *Frontiers in Aging Neuroscience* 5: 1–9.
Mark, V. H. et al. (1967). 'Role of Brain Disease in Riots and Urban Violence'. *JAMA* 201 (11): 895–896.
Mark. V. H. and F. R. Ervin (1970). *Violence and the Brain*. New York: Harper and Row.
Mayer, E. E. (1948). 'Prefrontal Lobotomy and the Courts'. *Journal of Criminal Law and Criminology* 38 (6): 576–583.
McMillan, J. (2014a). 'The Kindest Cut? Surgical Castration, Sex Offenders, and Coercive Offers'. *Journal of Medical Ethics* 40: 583–590.
McMillan, J. (2014b). 'Surgical Castration, Coercive Offers and Coercive Effects: It is Still Not About Consent'. *Journal of Medical Ethics* 40 (9): 596.
McTernan, E. (2018). 'Those Who Forget the Past: An Ethical Challenge from the History of Treating Deviance'. In D. Birks and T. Douglas (eds.) *Treatment for Crime*, pp. 274–288. Oxford: Oxford University Press.
Melo-Martin, I. and A. Salles (2014). 'Moral Bioenhancement: Much Ado About Nothing?'. *Bioethics* 29 (4): 223–232.
Monahan, J. and J. L. Skeem (2014). 'Risk Redux: The Resurgence of Risk Assessment in Criminal Sentencing'. *Federal Sentencing Report* 26: 391–435.
Monahan, J. and J. L. Skeem (2016). 'Risk Assessment in Criminal Sentencing'. *Annual Review of Clinical Psychology* 12: 489–513.
Montague, P. (1995). *Punishment as Societal Defence*. Lanham, MD: Rowman & Littlefield.
Moore, M. S. (1997). *Placing Blame*. Oxford: Clarendon Press.
Morris, N. (1974). *The Future of Imprisonment*. Chicago: Chicago University Press.
Morris, N. (1998). 'Incapacitation Within Limits'. In A. von Hirsch and A. Ashworth (eds.), *Principled Sentencing*, pp. 107–112. Oxford: Hart Publishing.
Morris, N. and M. Tonry (1990). *Between Prison and Probation*. New York: Oxford University Press.
Mosley, P. et al. (2014). 'Alienation and Authenticity in Parkinson's Disease and Its Treatment'. *AJOB Neuroethics* 5: 54–56.
Müller, S (2009). 'Body Integrity Identity Disorder (BIID)—Is the Amputation of Healthy Limbs Ethically Justified?'. *American Journal of Bioethics* 9 (1): 36–43.
Murphy, J. G. (1979). *Retribution, Justice, and Therapy*. Dordrecht: Reidel.
Nelson, L. and B. Ashby (2011). 'Rethinking the Ethics of Physician participation in Lethal Injection Execution'. *Hasting Center Report* 41: 28–37.

Nettle, D. et al. (2013). 'The Watching Eye Effect in the Dictator Game: It's Not How Much You Give, It's Being Seen to Give Something'. *Evolution and Human Behavior* 34 (1): 35–40.

Newman et al. (2010). 'Attention Moderates the Fearlessness of Psychopathic Offenders'. *Biological Psychiatry* 67: 66–70.

New York Times (1946) December 7. Quoted in 'Toward Rehabilitation of Criminals: Appraisal of Statutory Treatment of Mentally Disordered Recidivists'. *Yale Law Review* 57 (1948): 1085–1113.

Nozick, R. (1969). 'Coercion'. In S. Morgenbesser (ed.), *Philosophy, Science and Method*, pp. 440–472. St. Martin's Press.

Opton, E. M. (1974). 'Psychiatric Violence Against Prisoners: When Therapy is Punishment'. *Mississippi Law Journal* 45: 605–644.

Parfit, D. (1984). *Reasons and Persons*. Oxford: Clarendon Press.

Parfit, D. (2011). *On What Matters*. Oxford: Oxford University Press.

Pellegrino, A. (1993). 'Societal Duty and Moral Complicity'. *International Journal of Law and Psychiatry* 16: 371–391.

Pellegrino, E. (2001). 'The Internal Morality of Clinical Medicine: A Paradigm for the Ethics of the Helping and Healing Professions'. *Journal of Medicine and Philosophy* 26: 559–579.

Persson, I. and J. Savuslescu (2014). 'Against Fetishism About Egalitarianism and in Defense of Cautious Moral Bioenhancement'. *American Journal of Bioethics* 14 (4): 39–42.

Petersen, T. S. and K. Kragh (2017). 'Should Violent Offenders Be Forced to Undergo Neurotechnological Treatment? A Critical Discussion of the "Freedom of Thought" Objection'. *Journal of Medical Ethics* 43 (1).

Pfaff, J. F. (2017). *Locked In*. New York: Basic Books.

President's Council on Bioethics (2003). *Beyond Therapy: Biotechnology and the Pursuit of Happiness*. Washington DC: PCB.

Primoratz, I. (1990). *Justifying Legal Punishment*. London: Humanities Press International.

Pugh, J. (2017). 'Moral Bio-Enhancement, Freedom, Value and the Parity Principle'. *Topoi* (online first).

Pugh, J. and T. Douglas (2016a). 'Justifications for Non-Consensual Medical Intervention: From Infectious Disease Control to Criminal Rehabilitation'. *Criminal Justice Ethics* 35 (3): 205–229.

Pugh, J. and T. Douglas (2016b). 'Neurointerventions as Criminal Rehabilitation: An Ethics Review'. In J. Jacobs and J. Jackson (eds.), *The Routledge Handbook of Criminal Justice Ethics*, pp. 95–110. London: Routledge.

Pugh, J. and H. Maslen (2017). 'Drugs That Make You Feel Bad? Remorse-Based Mitigation and Neurointerventions'. *Criminal Law and Philosophy* 11 (3): 499–522.

Pugh, J. et al. (2017). 'Deep Brain Stimulation, Authenticity and Value'. *Cambridge Quarterly of Healthcare Ethics* 26: 640–657.

Pustilnik, A. C. (2009). 'Violence of the Brain: A Critique of Neuroscience in Criminal Law'. *Wake Forest Law Review* 44: 183–237.

Rafter, N. H. et al. (2016). *The Criminal Brain*. New York: New York University Press.

Ragon, S. A. (1995). 'A Doctor's Dilemma: Resolving the Conflict Between Physician Participation in Executions and the AMA's Code of Medical Ethics'. *University of Dayton Law Review* 20: 975–1007.

Rakic, V. (2014). 'Voluntary Moral Enhancement and the Survival-at-any-cost Bias'. *Journal of Medical Ethics* 40: 246–250.
Rakoff, J. S. (2016). 'Neuroscience and the Law: Don't Rush In'. *The New York Review of Books*, May 12: 30–32.
Ramachandran, G. (2009). 'Against the Right to Bodily Integrity: Of Cyborgs and Human Rights'. *Denver University Law Review* 87: 1–57.
Rao, R. P. N. et al. (2014). 'A Direct Brain-to-Brain Interface in Humans'. *Plos One* 9 (11): 1–12.
Raus, K. et al. (2014). 'On Defining Moral Enhancement: A Clarificatory Taxonomy'. Neuroethics 7: 263–273.
Raymond, M. J. (1956). 'Case of Fetishism Treated by Aversion Therapy'. *British Medical Journal* 2: 854–857.
Raynor, P. (2007). 'Community Penalties: Probation, What Works, and Offender Management'. In M. Maguire et al. (eds.), *The Oxford Handbook of Criminology*, pp. 1061–1099. Oxford: Oxford University Press.
Rice, M.E. and G. T. Harris (2011). 'Is Androgen Deprivation Therapy Effective in the Treatment of Sex Offenders?'. *Psychology, Public Policy, and Law* 17: 315–332.
Roberts, J. V. (2008). *Punishing Persistent Offenders*. Oxford: Oxford University Press.
Roberts, J. V. et al. (2003). *Penal Populism and Public Opinion*. New York: Oxford University Press.
Roberts, J. V. and A. Ashworth (2016). 'The Evolution of Sentencing Policy and Practice in England and Wales, 2003–2015'. *Crime and Justice* 45 (1): 307–358.
Roberts, J. V. and R. S. Frase (2019). 'The Problematic Role of Prior Record Enhancement in Predictive Sentencing'. In Keijser et al. (eds.), *Predictive Sentencing*, pp. 149–174. Oxford: Hart Publishing.
Roberts, J. V. and A. von Hirsch (ed.) (2010). *Previous Convictions at Sentencing*. Oxford: Hart Publishing.
Rocque, M. (2013). 'Experimental Neurocriminology: Etiology and Treatment'. In C. Welsh et al. (eds.), *Experimental Criminology*, pp. 43–64. Cambridge: Cambridge University Press.
Rose, N. (2010). 'Screen and intervene: Governing Risky Brains'. *History of the Human Sciences* 23 (1): 79–105.
Roskies, A. 2007. 'Neuroethics for the New Millennium'. In W. Glannon (ed.), *Defining Right and Wrong in Brian Science*, pp. 12–18. New York: Dana Press.
Roskies, A. 2016. 'Neuroethics'. In E. N. Zalta (ed.), *Stanford Encyclopedia of Philosophy* (online: https://plato.stanford.edu/).
Rothman, D. J. (1975). 'Behavior Modification in Total Institutions'. *The Hasting Center Report* 5 (1): 17–24.
Rothman D. J. (1995). 'Physicians and the Death Penalty'. *Journal of Law and Policy* 4: 151–160.
Ryberg, J. (2004). *The Ethics of Proportionate Punishment. A Critical Investigation*. Dordrecht: Kluwer Academic Publishers.
Ryberg, J. (2008). 'Moral Rights and the Problem of Privacy in Public: A Reply to Lever and Goold'. *Res Publica* 14: 49–56.
Ryberg, J. (2010). 'Recidivism, Retributivism, and the Lapse Theory of Previous Convictions'. In J. V. Roberts and A. von Hirsch (ed.), *Previous Convictions at Sentencing*, pp. 37–48. Oxford: Hart Publishing.

Ryberg, J. (2015a). 'Predictive Brain Devices, Therapeutic Activation Systems, and Aggression'. *AJOB Neuroscience* 6 (4): 36–38.
Ryberg, J. (2015b). 'Is Coercive Treatment of Offenders Morally Acceptable? On the Deficiency of the Debate". *Criminal Law and Philosophy* 9: 619–631.
Ryberg, J. (2016a). 'Compulsory Medication, Trial Competence, and Penal Theory'. *Law, Ethics and Philosophy* 4: 61–82.
Ryberg, J. (2016b). 'Deep Brain Stimulation, Psychopaths, and Punishment'. *AJOB Neuroscience* 7 (3): 168–169.
Ryberg, J. (2017). 'Neuroscience, Mind Reading, and Mental Privacy'. *Res Publica* 23: 197–211.
Ryberg, J. (2018). 'Neuroscientific Treatment of Criminals and Penal Theory'. In D. Birks and T. Douglas (eds.), *Treatment for Crime*, pp. 177–195. Oxford: Oxford University Press.
Ryberg, J. (2019). 'Risk and Retribution: On the Possibility of Reconciling Considerations of Dangerousness and Desert'. In J. de Keijser et al. (eds.), *Predictive Sentencing*, pp. 69–88. Oxford: Hart Publishing.
Ryberg, J. and A. Corlett (2010). *Punishment and Ethics: New Waves*. Basingstoke: Palgrave MacMillan.
Ryberg, J. and T. S. Petersen (2013). 'Neurotechnological Behavioural Treatment of Criminal Offenders—A Comment of Bomann-Larsen'. *Neuroethics* 6 (1): 79–83.
Ryberg, J. and T. Petersen (2014). 'Surgical Castration, Coercion, and Ethics'. *Journal of Medical Ethics* 40 (9).
Ryberg, J. and T. S. Petersen (2017). 'Neuroethics and Criminal Justice'. In K. Lippert-Rasmussen, K. Brownlee, and D. Coady (eds.), *A Companion to Applied Philosophy*, pp. 370–382. Oxford: Wiley.
Savulezcu, J. and I. Persson (2008). 'The Perils of Cognitive Enhancement and the Urgent Imperative to Enhance the Moral Character of Humanity'. *Journal of Applied Philosophy* 25 (3): 162–177.
Savulescu, J. et al. (2014). 'Response to Commentaries". In A. Akabayashi (ed.), *The Future of Bioethics: International Dialogues*, pp. 170–182. Oxford: Oxford University Press.
Savulescu, J. and I. Persson (2012). 'Moral Enhancement, Freedom and the God Machine'. *Monist* 95 (3): 399–421.
Scheid, D. E. (1997). 'Constructing a Theory of Punishment, Desert, and the Distribution of Punishments'. *The Canadian Journal of Law and Jurisprudence* 10: 441–506.
Schirmann. F. (2013). 'Badness, Madness and the Brain—the Late 19th Century Controversy on Immoral Persons and Their Malfunctioning Brains'. *History of the Human Science* 26 (2): 33–50.
Schnall, P. et al. (2008). 'Disgust as Embodied Moral Judgment'. *Personal and Social Psychology Bulletin* 34: 1096–1109.
Schnittker, J. (2014). 'The Psychological Dimensions and the Social Consequences of Incarceration'. *The Annals of the American Academy of Political and Social Science* 651 (1): 122–138.
Schoeman, F. (1981). 'On Incapacitating the Dangerous'. In H. Gross and A. Ashworth (eds.), *Sentencing*, pp. 39–45. New York: Oxford University Press.
Schoenthaler S. et al. (1997). 'The Effect of Randomized Vitamin-Mineral Supplementation on Violent and Non-violent Antisocial Behavior Among Incarcerated Juveniles'. *Journal of Nutritional and Environmental Medicine* 7: 343–352.

Scott, C. L. and T. Holmberg (2003). 'Castration of Sex Offenders: Prisoners' Rights Versus Public Safety'. *The Journal of the American Academy of Psychiatry and the Law* 31: 502–509.

Scull, A. (2015). 'Contending Professions: Science of the Brain and Mind in the United States, 1850–2013'. *Science in Context* 28 (1): 131–161.

Seddon, T (2007). 'Coerced Drug Treatment in the Criminal Justice System: Conceptual, Ethical, and Criminological Issues'. *Criminology and Criminal Justice* 7 (3): 269–286.

Senior, J. and J. Shaw (2007). 'Prison Healthcare'. In Y. Jewkes (ed.), *Handbook on Prisons*, pp. 376–398. Portland: Willan Publishing.

Serber, M. et al. (1975). 'Behavior Modification in Maximum Security Settings: One Hospital's Experience'. *American Crime Law Review* 85: 85–99.

Shaw, E. (2014). 'Direct Brain Interventions and Responsibility Enhancement'. *Criminal law and Philosophy* 8: 1–20.

Shaw, E. (2016). 'The Right to Bodily Integrity and the Rehabilitation of Offenders Through Medical Interventions: A Reply to Thomas Douglas'. *Neuroethics* (online first).

Shaw, E. (2018). 'Retributivism and the Biomedical Moral Enhancement of Offenders Through Brain Interventions'. In M. Hauskeller and L. Coyne (eds.), *Moral Enhancement: Critical Perspectives*, pp. 251–270. Cambridge: Cambridge University Press.

Shen, F. X. (2016). 'The Overlooked History of Neurolaw'. *Fordham Law Review* 85: 667–695.

Shniderman, A. B. and L. B. Solberg (2015). 'Cosmetic Psychopharmacology for Prisoners: Reducing Crime and Recidivism Though Cognitive Intervention'. *Neuroethics* 8: 315–326.

Shuster, E. (1997). 'Fifty Years Later: The Significance of the Nuremberg Code'. *The New England Journal of Medicine* 337: 1436–1440.

Singer, G. (1979). *Just Deserts*. New York: Balling Publishing Company.

Sio, P. S. de et al. (2012). 'The Necessity of Objective Standards for Moral Enhancement'. *AJOB Neuroethics* 3: 15–16.

Sitnick, S. A. (1975). 'Major Tranquilizers in Prison: Drug Therapy and the Unconsenting Inmate'. *Willamette Law Journal* 11: 378–397.

Sobchack, V. (2010). 'Living a "Phantom Limb": On the Phenomenology of Bodily Integrity'. *Body and Society* 16: 51–67.

Sparrow, R. (2014a). '(Im)moral Technology? Thought Experiments in the Future of Mind Control'. In A. Akabayashi (ed.), *The Future of Bioethics: International Dialogues*, pp. 133–119. Oxford: Oxford University Press.

Sparrow, R. (2014b). 'Egalitarianism and Moral Bioenhancement'. *American Journal of Bioethics* 14 (4): 20–28.

Sparrow, R. (2014c). 'Better Living Through Chemistry? A Reply to Savulescu and Persson on "Moral Enhancement"'. *Journal of Applied Philosophy* 31 (1): 23–32.

Spece, R. G. (1972). 'Conditioning and Other Technologies Used to Treat Rehabilitate Demolish Prisoners and Mental Patients'. *Southern California Law Review* 45: 616–684.

Sterling P. (1978). 'Ethics and Effectiveness of Psychosurgery'. In B. Brady (ed.), *Controversy in Psychiatry*, pp. 126–161. Philadelphia: Saunders Co.

Stinneford, J. F. (2006). 'Incapacitation Through Maiming: Chemical Castration, the Eighth Amendment, and the denial of Human Dignity'. *University of St. Thomas Law Journal* 3: 559–599.

Sugie, N. F. and K. Turney (2017). 'Beyond Incarceration: Criminal Justice Contact and Mental Health'. *American Sociological Review* 82 (4): 719–743.

Swayze, V. W. (1995). 'Frontal Leukotomy and Related Psychosurgical Procedures in the Era before Antipsychotics (1935–1954)'. *American Journal of Psychiatry* 152 (4): 505–515.

Tamburrini, C and J. Ryberg (ed.) (2012). *Recidivist Punishments*. Lanham: Lexington Books.

Ten. C. L. (1987). *Crime, Guilt, and Punishment*. Oxford: Clarendon Press.

Tokens, R. (2015). 'My Child Will Never Initiate Ultimate Harm: An Argument Against Moral Enhancement'. Journal of Medical Ethics 41: 245–251.

Tonry, M. (2001). 'Unthought Thoughts'. *Punishment and Society 3* (1): 167–181.

Tonry, M. (2004). *Punishment and Politics*. Devon: Willan Publishing.

Tonry, M. (2004). *Thinking About Crime*. New York: Oxford University Press.

Tonry, M. (2007). 'Determinants of Penal Policies'. Crime and Justice 36 (1): 1–48.

Tonry, M. (2009). 'Explanations of American Punishment Policies'. *Punishment and Society* 11 (3): 377–394.

Tonry, M. (2010). 'The Questionable Relevance of Previous Conviction to Punishments for Later Crimes'. In J. V. Roberts and A. von Hirsch (eds.), *Previous Convictions at Sentencing*. Oxford: Hart Publishing.

Tonry, M. (2016). 'Making American Sentencing Just, Humane, and Effective'. *Crime and Justice* 46 (1): 441–504.

Tonry, M. (2019). 'An Honest Woman's Guide to Punishment and Prediction'. In Keijser et al. (eds.), *Predictive Sentencing*, pp. 269–298. Oxford: Hart Publishing.

Torres, C. V. et al. (2013). 'Long-term Results of Posteromedial Hypothalamic Deep Brain Stimulation for Patients With resistant Aggressiveness'. *Journal of Neurosurgery* 119: 277–287.

'Toward Rehabilitation of Criminals: Appraisal of Statutory Treatment of Mentally Disordered Recidivists'. *Yale Law Review* 57 (1948): 1085–1113.

Tse, W. S. and A. J. Bond (2002). 'Serotonergic Intervention Affects Both Social Dominance and Affiliative Behaviour'. *Psychopharmacology* 161: 3234–330.

Tudor, S. K. (2008). 'Why Should Remorse be a Mitigating Factor in Sentencing?'. *Criminal Law and Philosophy* 2: 241–257.

Turney, K. et al. (2012). 'As Fathers and Felons: Explaining the Effects of Current and Recent Incarceration on Major Depression'. *Journal of Health and Social Behavior* 53 (4): 465–481.

Tyrer, P. et al. (2010). 'The Success and Failures of the DSPD Experiment: The Assessment and Management of Severe Personality Disorder'. *Medicine, Science and the Law* 50: 95–99.

Valdman, M. (2009). 'A Theory of Wrongful Exploitation'. *Philosophers' Imprint* 9 (6): 1–14.

Vanderzyl, K. (1994). 'Castration as an Alternative to Incarceration: An Important Approach to the Punishment of Sex Offenders'. *Northern Illinois University Law Review* 15: 107–159.

Vaske, J. C. (2017). 'Using Biosocial Criminology to Understand and Improve Treatment Outcomes'. *Criminal Justice and Behavior* 44: 1050–1072.

Veatch, R. M. and C. C. Macpherson (2010). 'Medical School Oath-Taking: The Moral Controversy'. *The Journal of Clinical Ethics* 21: 335–345.

Vincent, N. (2014). 'Restoring Responsibility: Promoting Justice. Therapy and Reform Through Direct Brain Interventions'. *Criminal Law and Philosophy* 8: 21–42.

von Hirsch, A (1993). *Censure and Sanctions*. Oxford: Clarendon Press.

Von Hirsch, A. (2010). 'Proportionality and the Progressive Loss of Mitigation: Some Further Reflections'. In J. V. Roberts and A. von Hirsch (ed.), *Previous Convictions at Sentencing*, pp. 1–16. Oxford: Hart Publishing.

von Hirsch, A. and A. Ashworth (eds.) (1998). *Principled Sentencing*. Oxford: Hart Publishing.

von Hirsch, A. and A. Ashworth (2005). *Proportionate Sentencing*. Oxford: Oxford University Press.

Von Hirsch, A. et al. (2009). *Principled Sentencing*. Oxford: Hart Publishing.

Von Hirsch, A, M. Wasik and J. Green (1989). 'Punishments in the Community and the Principles of Desert'. *Rutgers Law Journal* 20 (3): 595–618.

Walker, R. L. (2017). 'Dr. Death? Professionalism, Virtue, and U.S. Physician Participation in the Death Penalty'. *Criminal Justice Ethics* (online first): 1–19.

Wasserman, D. (2014). 'When Bad People Do Good Things: Will Moral Enhancement Make The Worlds a Better Place?'. *Journal of Medical Ethics* 40: 374.

Wertheimer, A. (1987). *Coercion*. Princeton: Princeton University Press.

Wertheimer, A. (1996). *Exploitation*. Princeton: Princeton University Press.

Wertheimer, A. and F. G. Miller (2014). 'There are (Still) no Coercive Offers'. *Journal of Medical Ethics* 40 (9): 592–593.

Wiegand, T. J. (2007). 'Captive Subjects: Pharmaceutical Testing and Prisoners'. *Journal of Medical Toxicology* 3 (1): 37–39.

Wilkins, L. T. (1975). 'Treatment on Trial'. *The Hasting Center Report* 5 (1): 35–48.

Wiseman, H. (2014). 'Moral Enhancement—Hard and Soft Forms'. *American Journal of Bioethics* 14 (4): 48–49.

Witt, K. et al. (2013). 'Deep Brain Stimulation and the Search for Identity'. *Neuroethics* 6: 499–511.

WMA (2008). https://www.wma.net/policies-post/wma-resolution-on-physician-participation-in-capital-punishment/

Woollett, K. and E. A. Maguire (2011). 'Acquiring the "the Knowledge" of London's Layout Drives Structural Brain Changes'. *Current Biology* 21 (24): 2109–2114.

Yang, Y. and A. Raine (2009). 'Prefrontal Structural and Functional Brain Imaging Findings in Antisocial, Violent, and Psychopathic Individuals: A Meta-Analysis'. *Psychiatry Research: Neuroimaging* 174 (2): 81–88.

Zaalberg, A. et al. (2010). 'Effects of Nutritional Supplements on Aggression, Rule-breaking, and psychopathology among young adult Prisoners'. *Aggressive Behavior* 36: 117–126.

Zellick, G. (1978). 'Corporal Punishment in the Isle of Man'. *International and Comparative Law Quarterly* 27: 665–671.

Zimmerman, D. (1981). 'Coercive Wage Offers". *Philosophy and Public Affairs* 10 (2): 121–145.

Zimring, F. E. (2001). 'Imprisonment Rates and the New Politics of Criminal Punishment'. *Punishment and Society* 3 (1): 161–166.

Zwolinski, M. and A. Wertheimer (2016). 'Exploitation'. In E. Zelta (ed.), *Stanford Encyclopedia of Philosophy* (online: https://plato.stanford.edu/).

Index

For the benefit of digital users, indexed terms that span two pages (e.g., 52–53) may, on occasion, appear on only one of those pages.

www.ingramcontent.com/pod-product-compliance
Ingram Content Group UK Ltd.
Pitfield, Milton Keynes, MK11 3LW, UK
UKHW020923170926
475595UK00031B/43

9 780190 846428